# Nanostructure Science and Technology

**Series editor**
David J. Lockwood, FRSC
National Research Council of Canada
Ottawa, Ontario, Canada



More information about this series at http://www.springer.com/series/6331

Stephen M. Goodnick • Anatoli Korkin
Robert Nemanich
Editors

# Semiconductor Nanotechnology

## Advances in Information and Energy Processing and Storage

Springer

*Editors*
Stephen M. Goodnick
Engineering Research Center
Arizona State University
Tempe, AZ, USA

Anatoli Korkin
Nano and Giga Solutions, Inc
St Paul, MN, USA

Robert Nemanich
Physics Department
Arizona State University
Tempe, AZ, USA

ISSN 1571-5744 ISSN 2197-7976 (electronic)
Nanostructure Science and Technology
ISBN 978-3-030-06324-5 ISBN 978-3-319-91896-9 (eBook)
https://doi.org/10.1007/978-3-319-91896-9


Softcover re-print of the Hardcover 1st edition 2018


Printed on acid-free paper

This Springer imprint is published by the registered company Springer International Publishing AG part of Springer Nature.
The registered company address is: Gewerbestrasse 11, 6330 Cham, Switzerland

# Preface

Energy and information are essential, and interconnected, elements for the development of human society. Transmission, processing, and storage of information requires energy consumption, while the efficient use and access to new energy sources requires new information (ideas and expertise) and the design of novel systems such as photovoltaic devices, fuel cells, and batteries. Semiconductor physics provides the knowledge base for the development of information (computers, cell phones, etc.) and energy (photovoltaics) technologies. The exchange of ideas and expertise between these two technologies is critical and expands beyond semiconductor nanotechnology.

This book is based on the tutorial lectures and research overviews presented at the Nano and Giga Challenges (NGC) 2017 conference in Tomsk (Russia), and is dedicated to solving scientific and technological problems in several areas of electronics, photonics, and renewable energy (photovoltaics). Progress in information and renewable energy technologies requires miniaturization of devices and reduction of costs, energy, and material consumption. The latest generation of electronic devices has nanometer-scale dimensions; new materials are being introduced into electronics manufacturing at an unprecedented rate, and alternative technologies to mainstream CMOS are evolving. Nanotechnology is widely accepted as a source of potential solutions in securing future progress for information and energy technologies.

The NGC conference series has had a long tradition of tutorial lectures given by world-renowned researchers. As early as the first forum in Moscow, Russia, in 2002, the organizers realized that publication of the lecture notes would be a valuable legacy of the meeting and a significant educational resource and knowledge base for students, young researchers, and experts alike. Our first book was published by *Elsevier* and named after the meeting itself – *Nano and Giga Challenges in Micro-*

*electronics* [1]. Our subsequent books based on the tutorial lectures of the NGCM2004 [2], NGC2007 [3], NGC2009 [4], NGC2011 [5], and NGC2014 [6] and the current book derived from the NGC2017 conference have been published by Springer (Springer Nature) in the *Nanostructure Science and Technology* series.

Arizona State University, Tempe, AZ, USA

Stephen M. Goodnick
Anatoli Korkin
Robert Nemanich

## References

1. Greer J, Korkin A, Labanowski J (eds) (2003) Nano and Giga challenges in microelectronics. Elsevier, Amsterdam
2. Korkin A, Gusev E, Labanowski J, Luryi S (eds) (2007) Nanotechnology for electronic materials and devices. Springer, New York
3. Korkin A, Rosei F (eds) (2008) Nanoelectronics and photonics: from atoms to materials, devices, and architectures. Springer, New York
4. Korkin A, Krstic P, Wells J (eds) (2010) Nanotechnology for electronics, photonics, and renewable energy. Springer, New York
5. Korkin A, Lockwood DJ (eds) (2013) Nanoscale applications for information and energy systems. Springer, New York
6. Goodnick S, Korkin A, Nemanich R (eds) (2015) Nanoscale materials and devices for electronics, photonics, and solar energy. Springer, New York

# Contents

# Contributors


**Gennadi Bersuker** Aerospace Corporation, El Segundo, CA, USA

**Vladimir Burtman** University of Utah, Salt Lake City, Utah, USA

**Holly Clingan** School of Molecular Sciences, Arizona State University, Tempe, AZ, USA

**Charles Dančak** VLSI Engineering, UCSC Silicon Valley Extension, Santa Clara, CA, USA

**Konstantin M. Degtyarenko** Siberian Physics and Technical Institute, Tomsk State University, Tomsk, Russia

**Igor Dorofeev** Department of Radioelectronics, Tomsk State University, Tomsk, Russia

**Grigory Dunaevsky** Department of Radioelectronics, Tomsk State University, Tomsk, Russia

**Ruslan M. Gadirov** Siberian Physics and Technical Institute, Tomsk State University, Tomsk, Russia

**Antonio García** Ira A. Fulton School of Engineering, Arizona State University, Tempe, AZ, USA

**David C. Gilmer** Nantero, Austin, TX, USA

**Stephen M. Goodnick** School of Electrical, Computer and Energy Engineering, Arizona State University, Tempe, AZ, USA

**Stanislav S. Khludkov** Functional Electronics Laboratory, Tomsk State University, Tomsk, Russia

**Tatjana N. Kopylova** Siberian Physics and Technical Institute, Tomsk State University, Tomsk, Russia

**Alex Laidlaw** School of Molecular Sciences, Arizona State University, Tempe, AZ, USA

**Vladimiro Mujica** School of Molecular Sciences, Arizona State University, Tempe, AZ, USA

**Sergey Yu. Nikonov** Siberian Physics and Technical Institute, Tomsk State University, Tomsk, Russia

**Ilya A. Prudaev** Functional Electronics Laboratory, Tomsk State University, Tomsk, Russia

**Pilarisetty Tarakeshwar** School of Molecular Sciences, Arizona State University, Tempe, AZ, USA

**Evgeny N. Telminov** Siberian Physics and Technical Institute, Tomsk State University, Tomsk, Russia

**Oleg P. Tolbanov** Functional Electronics Laboratory, Tomsk State University, Tomsk, Russia

**Micah Wimmer** School of Molecular Sciences, Arizona State University, Tempe, AZ, USA

**Vladimir P. Yakubov** Radio Physics Department, Tomsk State University, Tomsk, Russia

# Chapter 1
# Nanotechnology Pathways to Next-Generation Photovoltaics

**Stephen M. Goodnick**

**Abstract** In this book chapter, an overview is given of the latest advances and central challenges in photovoltaics research, and the role of nanotechnology in improving performance. Over the long term, nanotechnology is expected to enable improvements throughout the energy sector, but the most striking near- to midterm opportunities may be in lower-cost, higher-efficiency conversion of sunlight to electric power. Nanostructures in solar cells have multiple approaches by which they can improve photovoltaic performance: (1) new physical approaches in order to reach thermodynamic limits, (2) allow solar cells to more closely approximate their material-dependent thermodynamic limits, and (3) provide new routes for low-cost fabrication by self-assembly or design of new materials. We focus primarily on the first two approaches which have the goal of increasing efficiency. The limits of solar cell efficiencies are discussed, and several different approaches are described that circumvent long-held physical assumptions and lead beyond first- and second-generation solar cell technologies. The role of nanotechnology in specific cell technologies is reviewed, including its role in improving light-trapping and the light collection properties of solar cells, as well as dye-sensitized solar cells and perovskite solar cells, and recent advances in nanowire solar cells. Special emphasis is given on novel nanostructure-based devices based on advanced concepts such as hot-carrier cells, and multiexciton generation, which have the theoretical basis to realize high-efficiency energy conversion.

---

S. M. Goodnick (✉)
School of Electrical, Computer and Energy Engineering, Arizona State University, Tempe, AZ, USA
e-mail: stephen.goodnick@asu.edu





S. M. Goodnick et al. (eds.), *Semiconductor Nanotechnology*, Nanostructure Science and Technology, https://doi.org/10.1007/978-3-319-91896-9_1

## 1.1 Overview of Photovoltaics

### *1.1.1 History and Basic Principles*

Photovoltaic energy conversion is the direct conversion of light into electrical energy, without any intermediate steps such as steam generation in the case of solar thermal systems, for example. While photovoltaics is primarily associated with the revolution in semiconductor electronics in the latter half of the twentieth century, the photovoltaic effect itself was developed in 1839 by Edmond Bequerel in electrochemical cells. Although proceeded by earlier work, the first practical photovoltaics devices utilizing the solar spectrum, or solar cells, were demonstrated in 1954 at Bell Laboratories and found its first commercial applications as lightweight and long lifetime power sources for extraterrestrial applications for the emerging space industry in the late 1950s. Since that time, photovoltaics has developed into the fastest-growing source of terrestrial renewable energy, with an installed global capacity of over 400 GWs peak energy in 2017, and a production cost per watt that has dropped from $75/W in the late 1970s, to $0.30/W today, an exponential decrease in cost that is analogous to Moore's law for microelectronics.

The basic principal of operation of a photovoltaic device is illustrated in Fig. 1.1, which illustrates a pn junction under optical excitation. The operation of a photovoltaic device is truly quantum mechanical in nature, as quanta of electromagnetic radiation (photons) excite electrons from the valence band (filled states or orbitals) to empty or unoccupied states in the conduction band across a bandgap characteristic of the semiconductor (1.12 eV in the case of Si at room temperature, the dominant

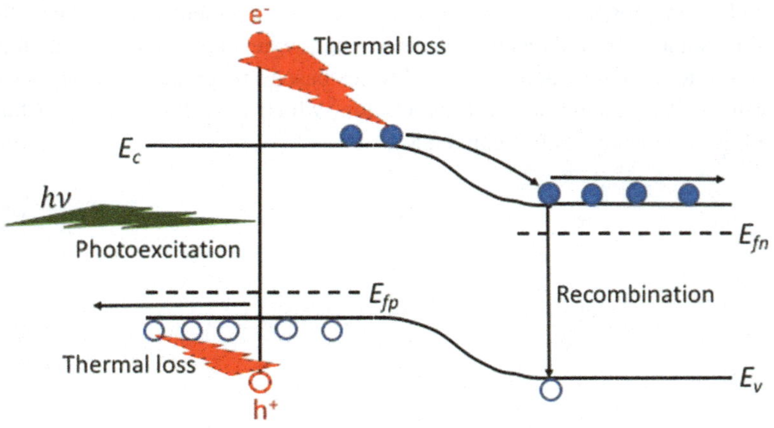


**Fig. 1.1** Band diagram of a pn junction solar cell illustrating the generation, thermalization, and capture of photoexcited carriers

semiconductor material). Photons with energy $h\upsilon$ below the bandgap are not absorbed since there are no final states available, whereas photons with energy above the bandgap create *electron-hole pairs* as shown. The excess kinetic energy of these photoexcited electrons and holes lows their energy to the crystal lattice of the material through electron-phonon (quanta of vibrational energy) interactions on very short time scales (femtoseconds to picoseconds) and relaxes to the minimum energy of the conduction band for electrons and the maximum energy of the valence band for holes. What is critical in performing useful work with such excitations is to have a basic asymmetry in the system that separates the electron and hole, provided in this case of the pn junction in Fig. 1.1 by the *space charge region* between the p- and n-regions, and high electric field there, which accelerates electrons to the right, and holes to the left, where they are collected in their respective n- and p-regions, giving rise to a photocurrent.

Solar cells for both terrestrial and space applications are optimized with respect to the broadband nature of the solar spectrum, shown in Fig. 1.2. The sun may be modeled, to a high degree of accuracy, as a blackbody source with intensity given by

$$I(\lambda) = R_{sun}^2/D^2 \times \frac{2\pi hc^2}{\lambda^5\left(\exp\left(\frac{hc}{\lambda kT}\right) - 1\right)}, \tag{1.1}$$

where $R_{sun}$ is the radius of the sun, $D$ is the distance from the earth to the sun, $T$ is the temperature of the sun, and $\lambda$ is the wavelength; the rest of the fundamental constants have their usual meaning. As can be seen in Fig. 1.2, the solar spectrum at the top of the atmosphere is well fit with Eq. (1.1) using a temperature of 5250 C (the more

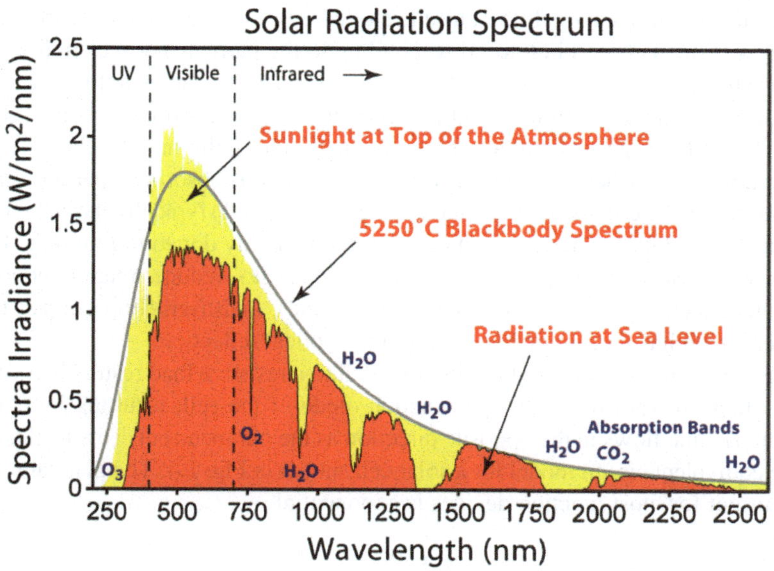


**Fig. 1.2** Spectral irradiance versus wavelength for the solar spectrum at the top of the atmosphere and at the earth's surface compared with the ideal blackbody spectrum (Wikipedia commons)

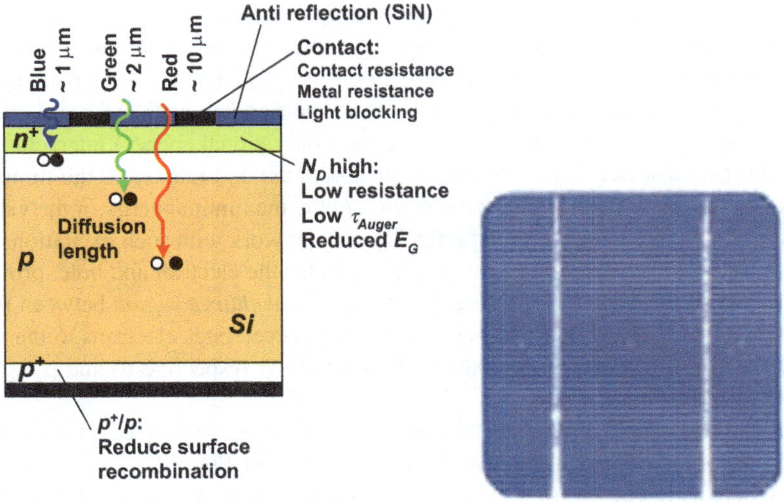


**Fig. 1.3** Cross-sectional schematic of a generic Si solar cell (*left panel*) and a top view of a solar cell (*right*) showing the grid pattern for allowing light absorption and carrier collection

accepted value is 5762 K), whereas due to absorption and reflection losses from atmospheric constituents (water, $CO_2$, etc.), as well as diffuse Raleigh scattering, the solar spectrum on the earth's surface is reduced with loss bands and is a function of latitude where the atmospheric path length increases for higher latitudes, further reducing the intensity.

The design of a solar cell therefore has to take into account not only the photon energy and intensity of photons corresponding to the particular wavelength of light but also the absorption coefficient of the material at a given wavelength, which is the inverse of the mean absorption depth of photons. This is illustrated in Fig. 1.3 for a basic Si cell design, where longer wavelength red light has a relatively long absorption depth, whereas short-wavelength blue light is absorbed near the surface. The structure of a commercial Si solar cell device (Fig. 1.1) typically has a thick base region (here p-type), a narrow emitter layer that is highly doped to minimize lateral resistance, heavy doping near the back contact to reduce recombination of photogenerated electrons there (back surface field), and a grid top contact to the emitter which has narrow fingers to minimize optical reflection.

The photocurrent delivered by the solar cell to a resistive load results in a voltage drop which forward biases the pn junction diode of the cell, resulting in a "dark" current, $I_D$, that flows in the opposite direction as the photocurrent. This is illustrated in the equivalent circuit model for a solar cell shown in Fig. 1.4. The general form of this current for most junction devices is exponential

$$I_D = I_0\left(e^{qV_D/nkT} - 1\right), \tag{1.2}$$

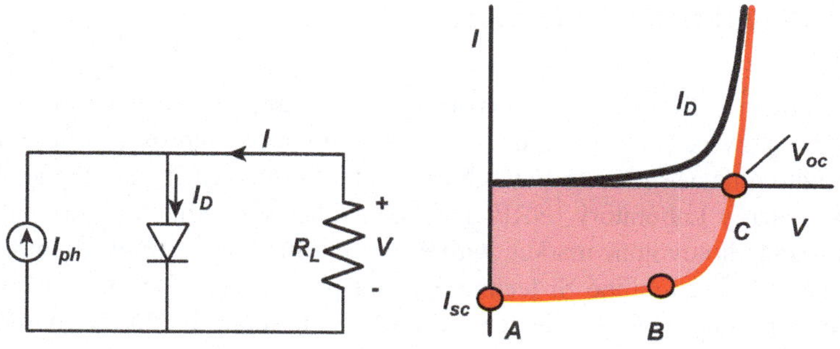


**Fig. 1.4** Equivalent circuit of the solar cell and the corresponding current-voltage characteristics and important points along the I–V curve

where $V_D$ is the voltage across the diode (which is different from $V$ in Fig. 1.4 if series resistance is included), $n$ is the ideality factor which varies between 1 and 2 usually, and $I_0$ is the reverse saturation current which depends explicitly on the recombination current in the junction, which can be radiative, due to traps or through Auger processes, and typically has an exponential dependence on the bandgap over the thermal voltage, i.e., $-E_g/kT$. As seen in terms of the net diode current, the I–V curve is shifted downward with light, corresponding to negative power or power generation. The points $A$, $B$, and $C$ along the curve denote the short-circuit current, the maximum power point, and open circuit, three figures of merit for photovoltaic devices. The short-circuit current is usually the same as the photocurrent induced by light absorption, $I_{ph}$, and is proportional to the photon flux above the bandgap incident on the surface (correcting for reflection and transmission through the cell). The open-circuit voltage depends on the reverse saturation current, $I_0$, and hence the bandgap of the material. Under open-circuit conditions (infinite $R_L$), $I_D = I_0$, so that using Eq. (1.2)

$$V_{oc} = \frac{nkT}{q} \ln\left(\frac{I_{sc}}{I_o} + 1\right). \tag{1.3}$$

Since $I_0$ decreases exponentially with the bandgap, $V_{oc}$ increases linearly with bangap, empirically given by $V_{oc} \approx E_G(in\ volts) - 0.4V$. Point $B$ corresponds to the maximum power point and is decreased from the maximum potential power $I_{sc}V_{oc}$, due to the "roundness" of the I–V curve, the degree of which is termed the *fill factor* (*FF*). In terms of the total optical energy incident on the device area, $P_{in}$, the optical to electrical conversion efficiency, $\eta$, is given by

$$\eta = \frac{V_{oc}I_{sc}FF}{P_{in}} \tag{1.4}$$

### 1.1.2 Photovoltaic Technologies

The performance of various photovoltaic technologies in terms of their solar to electrical energy conversion efficiency has continuously improved in a monotonic fashion with time as shown in Fig. 1.5 in a plot published each year by the National Renewal Energy Laboratory (NREL). Silicon solar cell technology dominates the current world photovoltaic market, with 95% controlled by a combination of single-crystal and multicrystalline Si technologies at present [1]. The highest efficiency reported for single-gap Si device technology to date is 26.7% based on a heterojunction structure using as crystalline Si substrate with thin layers of amorphous Si (a-Si), which forms a heterojunction due to the larger bandgap of a-Si (~1.7 eV) [2]. This cell record was demonstrated on a large area device, with an interdigitated backside contact which eliminates the front grid pattern of Fig. 1.3, increasing the photocurrent. Heterojunction AlGaAs/GaAs single-crystal solar cells have achieved even higher performance of 28.8%, the highest for any single-bandgap device [3]. Due to the cost and availability of material, such III–V semiconductor solar cells are normally too expensive for normal flat-plate solar except for space applications, where efficiency and radiation resistance are considerations.

The technologies above are wafer based, being comprised of either a wafer of the underlying material as the base region of the device or starting with a wafer and then removing the active region by lift-off to make thin solar cells. Thin film technologies are ones in which the active regions of the device are deposited using various thin-film methods onto a low-cost support material, for example, metal or glass. Such materials traditionally have lower efficiency but lower fabrication and materials costs, thus achieving lower $/W cost. Initially, amorphous Si (a-Si) thin-film solar was the dominant thin-film technology but has since been supplanted by II–VI CdTe heterostructure technology, which as shown in Fig. 1.5 has demonstrated over 21% performance by First Solar [4] and is the basis for several large (>200 MW) utility-scale solar installations worldwide. Recently, chalcogenide-based materials such as CIGS ($CuInGaSe_2$) have demonstrated similar high efficiencies and have taken an increasingly larger market share.

Organic thin-film solar cells are conceptually similar to the thin-film solar cells in the preceding paragraph, in which organic semiconductor materials are deposited onto glass or other support materials, however, defined by their HOMO (highest unoccupied molecular orbital) and LUMO (lowest unoccupied molecular orbital) levels. Rather than n- and p-type materials and the formation of a homojunction with a corresponding space charge region, a heterojunction is formed in which the lineup of the HOMO-LUMO levels of two different materials (called donor and acceptor materials) forms the junction. Due to the strong excitonic binding energy of organic materials, the photoexcited electron and hole are strongly bound as excitons, and diffuse to the heterointerface, where they dissociate due to the asymmetry of the barrier there into separate free electrons and holes on either side of the heterojunction, which are then collected at the contacts giving rise to a photocurrent. Due to the short diffusion lengths of excitons in such materials, *blended*

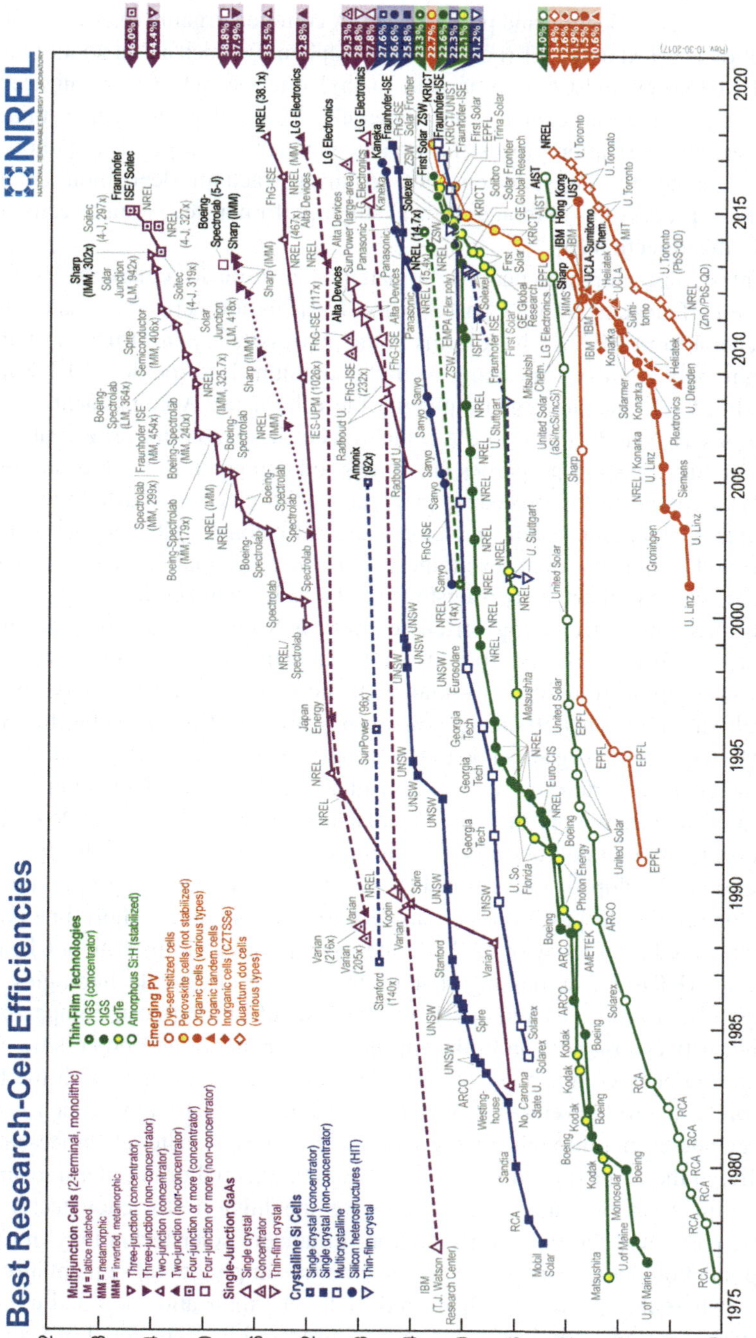


Fig. 1.5 Measured solar cell efficiency records for various technologies versus time (NREL https://www.nrel.gov/pv/)

*heterostructures* with the donor and accepter materials interdiffused into a network give better charge collection and performance. A common organic solar cell material combination is based on P3HT (poly(3-hexylthiophene)) which acts a donor material and PCBM (6,6-phenyl-C61butyric acid methyl ester) which acts as an acceptor material [5]. While efficiencies are generally lower than inorganic thin-film approaches, the fabrication process for organic is much less expensive, and organic layers can generally be spun on, not requiring vacuum deposition or high-temperature processing, although recent high-performance organic cells have moved in this direction [6].

Hybrid organic-inorganic perovskite materials have made remarkable advances in solar cell efficiencies over the past 5 years. These hybrid perovskite materials are of the general form $ABX_3$, which forms in the perovskite crystal structure, with the most commonly studied form being methylammonium lead iodide, $CH_3NH_3PbI_3$, with B=Pb, C=I, and the organic playing the role of the A component [7]. The bandgap of $CH_3NH_3PbI_3$ is 1.55 eV, which makes it suitable for terrestrial photovoltaics, and its transport properties are quite good compared to the typical organic materials discussed in the preceding paragraph, with electron and hole mobilities comparable to inorganic semiconductors, and diffusion lengths on the order of microns, with relatively weak excitonic effects. The original work on perovskite solar cells was based on a dye-sensitized solar cell architecture [8], which is a nanostructured device technology discussed more in Sect. 1.3.2, in which the hybrid perovskite is infused into a mesoporous wide-bandgap $TiO_2$ structure which acts as an electron acceptor from the perovskite, while holes are extracted in an electrolytic liquid cell structure. This efficiency was greatly improved by replacing the liquid electrolyte with a solid organic hole transport layer (spiro-MeOTAD), leading to the high efficiencies reported today in excess of 20%. A further innovation was the evolution of the mesoscopic structure to a planar structure in which $CH_3NH_3PbI_3$ is treated essentially as a polycrystalline semiconductor [9], although both approaches continue to be developed. As seen in Fig. 1.5, the performance of perovskite solar cells has one of the steepest slopes of any technology over a very short time period, reaching a record as of today of 22.7% at the Korean Research Institute of Chemical Technology (KRICT), surpassing that of thin-film inorganic technologies. The advantage of perovskite technology is the low manufacturing cost, comparable to something between organic and thin-film inorganic material and processing costs. The main barrier to commercialization to date are issues associated with the long-term stability due to the sensitivity of the perovskite to water vapor, which various groups are addressing through encapsulation and improved materials processing.

Finally, in the context of this brief review of commercial or near-term technologies, multijunction or tandem solar cells are the highest-efficiency technology presently, particularly at high concentration. In tandem solar technology, multiple bandgap junction devices are connected together or simply grown sequentially on a substrate, where the multiple bandgaps reduce the thermalization loss and transparency issues of a single-bandgap solar cell, as discussed in more detail in Sect. 1.2. The highest-efficiency tandem devices are single-crystal III–V materials grown epitaxially on top of one another, starting with the lowest-bandgap material, and ending with the highest bandgap, in terms of the direction of the incident radiation. In this way, the short-wavelength light is absorbed in the top wide-bandgap material, and the longer-wavelength light is absorbed by subsequent layers. The cells are typically connected together in series with tunnel junctions, such that overall, the current is the same through all the cells, and the overall cell voltage is the sum of the voltages of the individual cells. The cost of the substrates and high-quality epitaxial growth (using, e.g., molecular beam epitaxy, or MBE) makes the cost per cell quite high. The high cost of these devices is compensated for by using them in an optical concentrator (where the light intensity is 200×–400× higher than typical sunlight), such that only very small areas are needed. The optical systems must track the sun, and these large systems are suited primarily for utility-scale applications. Three junction cells based on Ge/GaAs/GaInP or similar combinations have exceeded 40% efficiency as shown in Fig. 1.5, and the record as of the time of this writing is 46% from Fraunhofer ISE and Soitec at a concentration of 297×.

## 1.2 Limits of Efficiency

### *1.2.1 Detailed Balance Analysis*

A photovoltaic device may be ideally analyzed independent of its material parameters (apart from bandgap) from thermodynamic considerations only, called *detailed balance*. Shockley and Queisser's 1961 paper [10] is based on an idealized description of a solar converter which includes no details of the cell structure itself; rather, it assumes complete collection of available photogenerated carriers with the following basic assumptions: (1) radiative recombination only, (2) one bandgap, (3) absorption across the bandgap in which one photon generates one electron-hole pair, (4) constant temperature in which the carrier temperature is equal to the lattice and ambient temperature, and (5) steady state, close to equilibrium. Mathematically, the current density is written in terms of three terms

$$J = qg\left[\left(fC\int_{E_G}^{\infty}\frac{E^2dE}{\exp\left(\frac{E}{kT_{sun}}\right)-1}+(1-fC)\int_{E_G}^{\infty}\frac{E^2dE}{\exp\left(\frac{E}{kT}\right)-1}\right)-\int_{E_G}^{\infty}\frac{E^2dE}{\exp\left(\frac{E-qV}{kT}\right)-1}\right], \tag{1.5}$$

with

$$g=\frac{2\pi}{h^3c^2}\quad f=\left(\frac{R_{sun}}{D_{sun}}\right)^2 \tag{1.6}$$

where $C$ is the concentration factor, $E_g$ is the material bandgap, and $V$ is the voltage across the cell. Equation (1.5) is written in terms of blackbody sources of photons, in which the first term represents the incident photon flux from Eq. (1.1), with $T_{sun}$ the

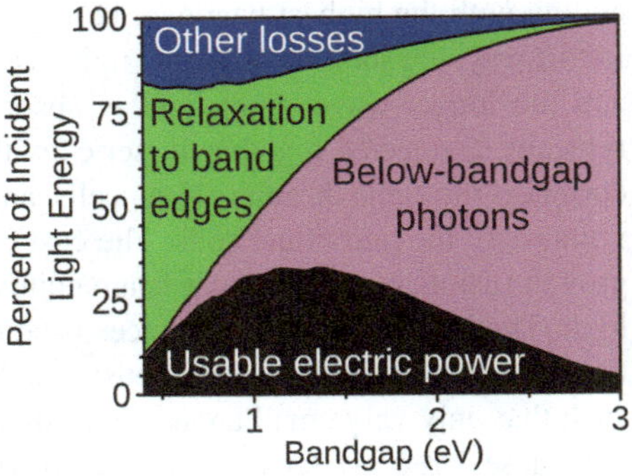


**Fig. 1.6** The Shockley-Queisser efficiency limit (black curve) versus bandgap for the AM1.5 solar spectrum and the contributions to this limit due to different loss mechanisms (Source: Wikipedia Commons)

temperature of the sun, and the second term represents the blackbody radiation from the surroundings at the local ambient temperature, $T$. The third term represents the blackbody radiation re-radiated by the absorber which is out of equilibrium in terms of the voltage $V$, which represents the splitting of the quasi-Fermi energies within the material. If a solar spectrum other than the ideal blackbody spectrum is used, then the first term is replaced simply by the integral of the photon flux above the bandgap.

In Eq. (1.6), the "dark" current discussed in the previous section is only due to blackbody radiation from the semiconductor absorber, which in effect is due to radiative recombination within the semiconductor generating photons above the bandgap. By evaluating the integrals for a range of voltages going from zero to the bandgap, the maximum $JV$ product is found, giving the conversion efficiency in terms of the total incident power. The result of this calculation is plotted in Fig. 1.6, where the black curve is the calculated efficiency versus bandgap for an AM1.5 terrestrial solar spectrum. The maximum efficiency without concentration is around 33.7% corresponding to maxima at 1.1 and 1.4 eV. The principal losses are due to the loss of photons with energy below the bandgap, and loss of the excess energy of the photon above the bandgap in terms of energy relaxation of photoexcited carriers back to the band edges, with thermalization being the main loss for small bandgap materials and optical transparency the main loss for high bandgaps.

### 1.2.2 Exceeding the Shockley-Queisser Limit

As discussed above, main factors contributing to Shockley-Queisser (SQ) limit in Fig. 1.6 are that photons below the bandgap of the absorber are not collected, while any excess kinetic energy in the electron-hole pair created by a photon above the bandgap is quickly lost through thermalization and therefore only contributes the energy an electron-hole pair at the bandgap, independent of the photon energy. The variation and limiting value then of the efficiency with bandgap is a direct result of the specific broadband nature of the solar spectrum, with a peak at 33.7%. In contrast, the theoretical limit of solar to electrical energy conversion has been considered by several authors based on thermodynamics alone and is approximately

85% [11], and therefore there is a substantial gap between the single-gap SQ limit and what should be possible.

There are various pathways to approaching thermodynamic conversion efficiencies rather than the single-gap SQ limit, by circumventing the assumptions inherent in the SQ analysis, which we discuss in more detail below.

**Broadband Solar Spectrum** As mentioned above the fact that the solar spectrum is a broad band source leads to the trade-offs between transparency to below bandgap photons and thermalization energy losses for those above. If the solar spectrum could be transformed to a narrower spectrum, higher-efficiency performance is possible. Up/down conversion of the solar spectrum through phosphors or two-photon absorption/emission are potential methods to accomplish this.

**Multiple Electron-Hole Pairs per Photon** The SQ analysis assumes a single electron-hole pair (EHP) excitation per photon, but the excess energy of the photon above the gap may be sufficient to produce a second or third, etc. EHP. Another route to exceeding the single-gap limit is to generate multiple electron-hole pairs from a single photon through the creation of secondary carriers. The process of impact ionization in semiconductors by high-energy charge carriers is well known, and the potential considered for photovoltaics [12]. More recently, nanostructured systems such as quantum dots and nanowires have shown particularly promising results due to quantum confinement effects, where the effect is often referred to as multiexciton generation (MEG), due to the importance of excitonic states in strongly confined systems, which we discuss in more detail in Sect. 1.3.5 [13].

**Extraction of Hot Carriers Before Thermalization** To circumvent the loss associated with thermalization (the loss of excess kinetic energy of photoexcited carriers), Ross and Nozik proposed the concept of hot-carrier solar cells [14]. In this concept, electrons and holes are not collected at the band edges (which limits to the output voltage to the bandgap), rather they are collected through energy-selective contacts above and below the conduction and valence band edges, respectively, effectively increasing the operating voltage. The absorber material suppresses energy loss, so that hot carriers can reach sufficient energy to escape through the energy-selective contacts. The concept was extended further by Würfel and coworkers who considered the effect of impact ionization and secondary carrier generation on the ultimate efficiency of this concept [15, 16]. Recent results are discussed in more detail later in Sect. 1.3.6.

**Multiple Bandgaps/Energy Levels** The SQ analysis considered only a single-gap material, but already in the mid-1950s, it was recognized that multijunction or tandem solar cells were capable of efficiencies above that of single-gap devices. Tandem solar cells have shown the highest efficiencies of any solar cell technology [17], with the record to date in excess of 46% in a four-junction structure [18], which greatly exceeds the single-gap SQ limit. The detailed balance approach given by Eq. (1.5) can be generalized to consider multiple junctions, each with its own detailed balance equation and with a modified spectrum according to the number of cells above or below (due to reabsorption of emitted light) the particular junction

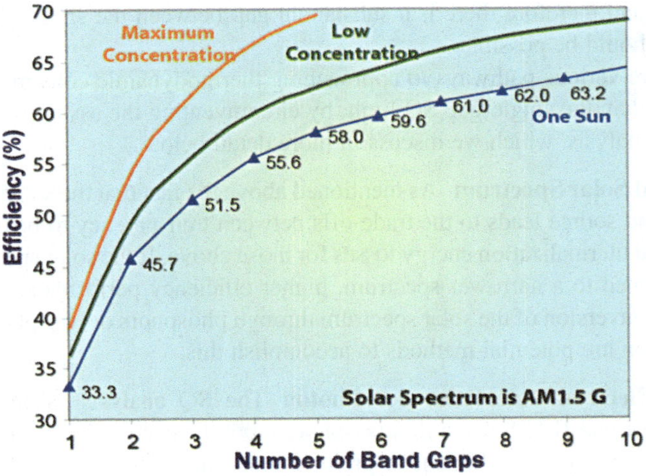


**Fig. 1.7** Calculated detailed balance efficiency of as a function of the bandgap number for a tandem cell, for three different concentrations of AM1.5 spectrum sunlight

in question. The results of this calculation for different numbers of bandgaps for normal, low, and high concentration are shown in Fig. 1.7. Concentration provides a significant improvement in performance for high number of junctions, which can be explained simplistically in terms of Eq. (1.3) for the open-circuit voltage increase with photocurrent (proportional to concentration) and having junctions in series where the effect is additive, multiplying the concentration effect. For maximum concentration (set by étendue limits to 46,050X), one can approach the thermodynamic limit with an arbitrarily large number of junctions.

As discussed earlier, commercial tandem cells are grown in series using epitaxial material growth technology, which is generally quite expensive compared to conventional Si solar cell manufacturing and has many material challenges to both optimize the bandgaps and have lattice-matched materials for low-defect growth. There has been a recent revival in interest in Si tandem solar cells, e.g., increasing the efficiency of current Si technology with an additional junction grown on a Si substrate, to improve the performance without a substantial cost increase. For a 1.12 eV lower-bandgap material, the optimum bandgap for a top material is 1.7 eV. Based on lattice-matching considerations, GaP is one of the few that is nearly lattice matched to S and can be an alternative to a-Si as a heterojunction technology [19]. Its 2.36 eV bandgap is unsuitable for monolithic tandem applications, which has led to consideration of dilute nitride materials in order to match both bandgap and lattice constant [20].

Rather than fabricating multiple junctions, another approach is to introduce multiple levels within the same material, which provide multiple paths for photon absorption but collect carriers at the primary bandgap of the host material. Luque and Marti introduced the concept of an intermediate-band (IB) solar cell to realize such a structure and overcome the SQ limit [21], while similar concepts had been suggested for quantum well solar cells. An intermediate level in the bandgap is introduced through, for example, self-assembled quantum dots, which allow low-energy photons to excite electron-hole pairs through multiphoton absorption, below the gap of the principal absorber.

## 1.3 Nanotechnology Pathways for Photovoltaics

Nanotechnology refers to technology at literally nanometer-scale dimensions ($10^{-9}$ meters), although the term is used somewhat loosely for devices with critical feature sizes below 100 nm, which is a broad umbrella encompassing a host of scientific and engineering disciplines including life sciences, physics, chemistry, engineering, and computer science, among others. Nanotechnology has been driven by remarkable advances in materials synthesis, nanofabrication, and atomic-scale characterization over more than four decades. The nanotechnology and nanoscience fields represent a convergence of many different disciplines, partly driven by top-down miniaturization driven by Moore's law and the microelectronics industry and bottom-up approaches driving chemistry and the life sciences, where self-ordered nanoscale structures are naturally occurring and responsible for the exquisite functionality that exits in biomolecular structures. The ability to visualize and characterize atomic-scale features began with remarkable advances in high-resolution electron microscopy and lattice imaging and then invention of the scanning tunneling microscopy and atomic force microscopes [22], which allows atomic-level imaging of atomic positions, spectroscopic features, and positioning of atoms on a surface.

Top-down nanofabrication techniques such as electron-beam, ion-beam, and deep ultraviolet (UV) lithography allow the patterning of features down to 10s of nanometers, and AFM techniques can be used to actually position atoms literally with atomic precision. At the same time, there have been significant advances in "bottom-up" synthesis and control of self-assembled materials such as nanoparticles, nanowires, molecular wires, and novel states of carbon such as fullerenes, graphene, carbon nanotubes, and composites thereof. These advances have led to an explosion of scientific breakthroughs in studying the unique electronic/optical/mechanical properties of these new classes of materials.

Nanostructures in solar cells have multiple approaches by which they can improve photovoltaic performance: (1) new physical approaches in order to reach thermodynamic limits, (2) allow solar cells to more closely approximate their material-dependent thermodynamic limits, and (3) provide new routes for low-cost fabrication by self-assembly or design of new materials. Some of the specific advantages and disadvantages presented by nanotechnology are listed below:

*Advantages*

- Range of bulk materials with proper energy gaps, catalytic properties, etc. very limited: Nanostructured materials allow "bandgap engineering" of electronic states and energy gaps: artificial materials.
- Provide intermediate energy centers within host material.
- Optical absorption can be increased; reflection and other optical losses decreased.
- Improve transport and reduce scattering and energy loss.

*Disadvantages*

- Higher surface-to-volume ratio means surface effects dominate: higher recombination.

In the following, we first summarize what the important nanomaterial technologies consist of in terms of nanoparticle, nanowires, and quantum wells. This is followed by consideration of different nanotechnology-based solar cell architectures such as dye-sensitized solar cells, and nanowire solar cells, as well as a discussion of light management in photovoltaic devices through nanostructures. We then end considering two advanced concept technologies including multiexciton generation devices and hot-carrier solar cells, followed by a summary.

### 1.3.1 Nanomaterials

Nanomaterials usually refers to materials that have structural features on the nanoscale and in particular their properties stem from these nanoscale dimensions. Such nanomaterials may include quantum wells, nanoparticles, nanopowders, nanoshells, nanowires, nanorods, nanotubes such as carbon nanotubes, nanomembranes, and nanocoatings or combinations of these to form nanocomposites. An important feature of nanomaterials for energy applications compared to their bulk counterparts is that the surface-to-volume ratio is greatly enhanced, resulting in fundamental changes in the chemical, electronic, mechanical, and optical properties, in essence creating a new material. Such changes are a result of the different energies associated with surfaces compared to the bulk. This may result in complete changes in the way materials may behave, in terms of their catalytic properties, their chemical bonding, strength, etc. Another effect is the so-called quantum size effect, which, like the simple particle in a box, quantizes the motion of electrons in a solid, meaning the allowed energies can only assume certain discrete values. This generally changes the electrical and optical properties of materials. For example, nanoparticles show a blue shift in their absorption spectrum to high frequency due to quantum confinement effects.

## Quantum Wells and Superlattices

One of the first truly nanoscale fabrication technologies was the development of precision epitaxial material growth techniques such as molecular beam epitaxy (MBE) [23] and metal-organic chemical vapor deposition (MOCVD), through which high-quality, lattice-matched heterojunction (junction between two dissimilar materials) semiconductor-layered systems could be realized, with atomic precision in the interface quality. A sandwich composed of a narrower bandgap material clad with larger bandgap materials of atomic dimensions is referred to as a quantum well (QW), and when many of these are grown sequentially, they are referred to as a multi-quantum well (MQW) system. These systems exhibit strong quantum confinement effects due to the low density of defects at the interface of lattice-matched materials such as GaAs and $Al_xGa_{1-x}As$. If the thickness of the barriers separating large and small bandgap materials is reduced so that the electronic states of the QWs overlap, the system is referred to as a superlattice (SL), which behaves as a new material electronically.

The capability of epitaxial growth to realize atomically precise heterointerfaces has served as the basis for a number of electronic and optoelectronic device technologies including heterojunction bipolar transistors (HBTs), high-electron mobility transistors (HEMTs), quantum well lasers, quantum well infrared photodetectors (QWIPs), and quantum cascade lasers (QCLs), to mention a few. In photovoltaic applications, single-crystal epitaxial growth is the basis for high-efficiency tandem or multijunction solar cells which hold the record for conversion efficiency as discussed earlier. They typically are designed for high-performance extraterrestrial applications (space-craft) or high-performance terrestrial concentrating photovoltaic (CPV) applications. MQW systems are also of active interest for QW solar cells or several of the advanced concept devices discussed in the next section.

## Nanowires

The term *nanowire* generally refers to a high aspect ratio wire-like structures in which the cross-sectional dimensions are nanometer scale, while the length may be micro- to macroscale. Nanowires are generally solid, not hollow structures, the latter being referred to as *nanotubes*. Such nanowires may be oxide, metallic, or semiconducting. One of the major broad techniques used for the growth of semiconducting nanowires is *vapor-phase synthesis*, in which nanowires are grown by starting from appropriate gaseous components. In the so-called *vapor-liquid-solid* (VLS) mechanism, which uses metallic nanoparticles as seed sites to stimulate the self-assembled growth of nanowires. The desired semiconductor system is introduced in terms of its gaseous components, and the entire assembly is heated to a temperature beyond the eutectic temperature of the metal/semiconductor system. Under these conditions, the metal forms a liquid droplet, with a typical size of a few nanometers. Once this droplet becomes supersaturated with semiconductor, it essentially nucleates the growth of the nanowire from the base of the droplet. Figure 1.6 shows examples of Si nanowires grown by VLS method using gold nanoparticles as the seeding droplets. The high-crystalline integrity of this nanowire can be clearly seen in this image, which also makes clear how the diameter of the nanowire is connected the size of the catalyst droplet [24]. The wire shown here was grown by using chemical vapor deposition (CVD) to generate the semiconductor precursors, a popular approach to VLS. Other methods may also be used, however, including laser ablation and MBE. The VLS method has emerged as an extremely popular method for the fabrication of a variety of nanowires. It has also been used to realize various III–V (GaN, GaAs, GaP, InP, InAs) and II–IV (ZnS, ZnSe, CdS, CdSe) semiconductor nanowires, as well as several different wide-bandgap oxides (ZnO, MgO, $SiO_2$, CdO).

Samuelsson and coworkers have also had enormous success in developing nanoscale electronic devices that utilize VLS-formed, III–V semiconductor, nanowires as their active elements [25]. They have demonstrated that heterostructure nanowires of InAs and InP, as well as GaAs and InAs, can be realized that have very sharp heterointerfaces [26]. They have subsequently used this technique to implement a variety of nanoscale devices, such as resonant-tunneling diodes [27] and single- [28] and multiple-coupled [29, 30] quantum dots. The strong lateral confinement generated in these structures, combined with their high-crystalline quality, endows them with robust quantum-transport characteristics. Quantum dots realized using these structures show very clear single-electron tunneling signatures, with evidence that the g-factor of the electrons can be tuned over a very wide range [31]. The ability to arbitrarily introduce serial heterointerfaces into such nanowires should offer huge potential in the future for the further development of novel nanodevices (Fig. 1.8).

From the perspective of energy conversion, nanowire structures are being researched as new materials for electrochemical storage and energy conversion devices, due to the large surface-to-volume ratio of these structures, which improves

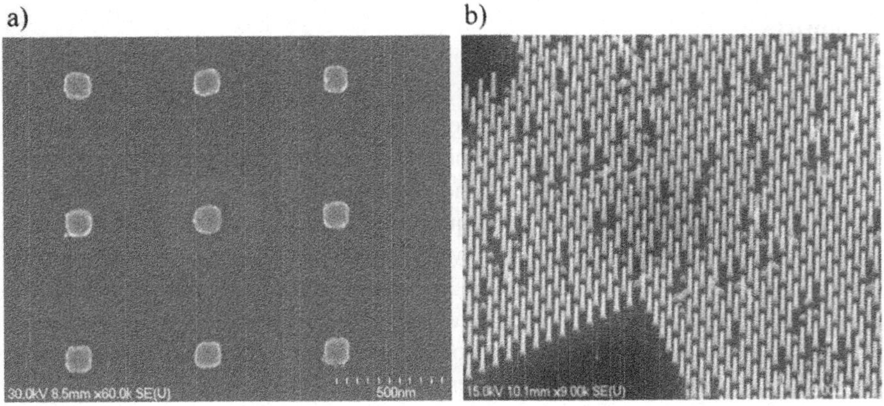


**Fig. 1.8** Self-assembled growth of nanowires using vapor-liquid-solid (VLS) epitaxy. (**a**) Scanning electron micrograph of Au seeds patterned with electron-beam lithography. (**b**) Ge nanowires after growth on Si (111) [32]

the catalytic performance and reaction rates, as well as provides large internal surface areas for charge storage. Within renewable energy technologies such as solar photovoltaic devices, nanowires are finding increasing use in light management, reducing the amount of light lost and allowing less material to be used for the absorption of light, hence improving efficiency and lowering material cost. Most of these efforts are in the research phase or as part of start-up ventures commercially.

## Nanoparticles and Quantum Dots

Nanoparticles is a name generally given to ultrafine size particles with dimensions on the order of 1–100 nm. If the nanoparticles are single-crystal individual particles, they are often referred to as nanocrystals [33]. Alternately, agglomerates of nanoparticles are referred to as nanopowders. Nanoparticles can be metals, dielectrics, or semiconductors. They can also be grown with different compositions to form core-shell nanoparticles with unique electrical and optical properties, as illustrated in Fig. 1.7. Their electronic and optical properties are different from bulk materials as mentioned before due to quantum size effects which shift the fundamental gap to higher energy. Surface effects also play a dominant role. In particular, the dielectric properties can also be modified by surface plasma resonance effects, which change the absorption properties. The high surface-to-volume ratio affects other properties such as diffusion properties in liquid and the adhesive properties.

Nanoparticles are synthesized by a variety of techniques. One inexpensive method is through ball mill micro-machining to literally grind materials down into nanoparticles. Pyrolysis and rf plasma techniques may also be used. A popular method for synthesizing nanoparticles of high quality is through chemical solution methods; in particular sol-gel methods can realize colloidal solutions of nanoparticles which may be subsequently dried for individual nanoparticles, or the gel solutions cast for particular applications. Another method of realizing semiconductor nanoparticles is through self-assembly of InAs or InGaAs quantum dots that on a GaAs substrate via the Stransky-Krastinov growth process [34]. In this mode of growth, a thin layer of InAs is grown on top of a GaAs substrate, but, if the layer is sufficiently thin, the strain will cause the InAs to agglomerate into small three-dimensional quantum dots.

Nanoparticles (and other nanomaterials such as nanowires and nanotubes) can be embedded in a host matrix to form a *nanocomposite*. The main differentiating factor between a nanocomposite and a normal composite material is the large surface-to-volume ratio of the nanoparticle, which means that there is a large internal surface area associated with the nanoparticles compared to normal composite materials. Therefore, a much smaller amount of nanoparticle composition can have a much greater effect on the overall nanocomposite properties. Nanocomposites can be comprised of many forms, the primary ones be ceramic matrix, metal matrix, or polymer matrix nanocomposites (Fig. 1.9).

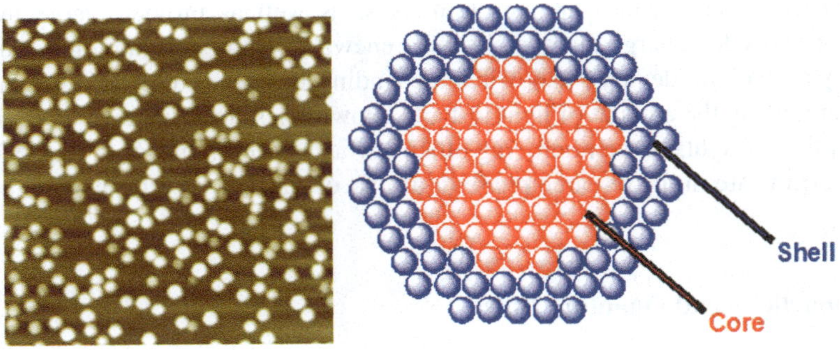


**Fig. 1.9** Micrograph of self-assembled InAs nanoparticles on a GaAs substrate (left) [35] and a schematic of a core-shell nanoparticle (wiki commons)

## *1.3.2 Dye-Sensitized Solar Cells*

Dye-sensitized solar cells (DSSC) are based on an electrochemical cell structure harkening back to Becquerel's experiments in the 1800s, which first demonstrated the photovoltaic effect. The DSSC was first realistically demonstrated by O-Regan and Grätzel in 1991 [36] and by its construction can be considered one of the first applications of nanotechnology to solar cells. Since this first demonstration, traditional DSSC efficiencies have reached over 11%. They have since been superseded by perovskite solar cell technology, as discussed in Sect. 1.1.2, as the first perovskite solar cells evolved from the DSSC architecture, and the highest efficiency perovskite cells still incorporate a nanostructured DSSC-like structure.

A schematic of a typical DSSC architecture is shown in Fig. 1.10. It consists of nanoparticle or nanostructured $TiO_2$, which is a wide-bandgap material ($E_g$=3.2 eV). It is usually formed on a glass substrate with a transparent conducting oxide (TCO) such as indium tin oxide (ITO) or fluorine-doped tin oxide (FTO), which is the side light enters. A dye material such as a ruthenium (Ru) is introduced through, for example, a liquid spin-on coat and dry process, which adheres the dye to the $TiO_2$ nanostructure, to realize large surface area coverage. The HOMO-LUMO separation of the dye is matched to the solar spectrum, and when light is absorbed by the dye, the excited $e^-$ transfers to $TiO_2$ as shown, due to the lineup of the conduction band of the $TiO_2$ relative LUMO level of the dye. The electron quickly diffuses to the TCO and the external circuit. The positive charge in the HOMO level reacts via a redox couple in the electrolyte; a typical electrolyte in DSSC cells is iodine based, consisting of $I^-$ and $I_3^-$. The dye is reduced and the iodine oxidized by a process in which $3I^- = 2e^- + I_3^-$, that is, two holes in the dye are neutralized by converting three iodine ions into one $I_3^-$ singly charged molecule, giving up to electrons to the dye. At cathode, the positive charge carrier, $I_3^-$ is reduced and converted back to $3I^-$ by the transfer of two electrons from the cathode, thus completing the circuit.

The improvement in efficiency over time of DSSCs initially increased rapidly and then plateaued somewhat as seen in Fig. 1.5. Some of the limitations of the liquid

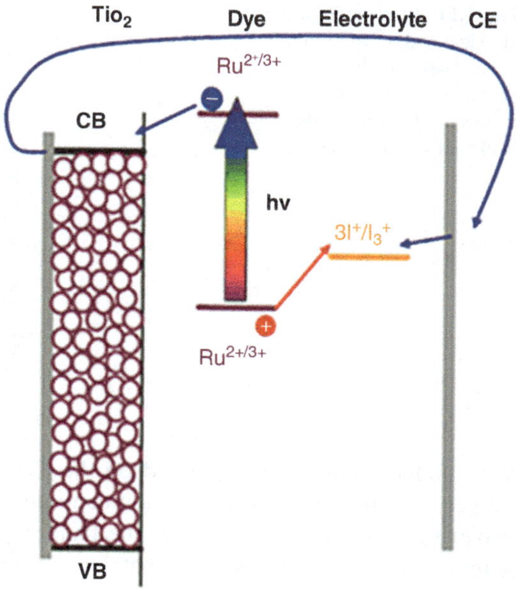


**Fig. 1.10** Schematic of dye-sensitized solar cell (DSSC) architectures (wiki commons)

electrolyte approach to DSSC have included stability issues associated with the liquid electrolyte itself, the relatively narrow spectral absorption in the dyes, and low open-circuit voltages to recombination processes such as at the dye-$TiO_2$ interface. Improvements in finding new dye materials including inorganic nanoparticles with broad absorption and improvements in the structure to reduce recombination processes have led to performance improvements in recent years.

The most dramatic evolution of the DSSC architecture was, as discussed earlier, the replacement of conventional dyes with the hybrid perovskite $CH_3NH_3PbI_3$, which is a semiconductor material with a bandgap of 1.55 eV and could be introduced into the nanoporous $TiO_2$ matrix through low-temperature processing. Further, the liquid electrolyte was replaced by a solid organic hole-transport layer (spiro-MeOTAD) with good transport properties, leading to a much more compact planer geometry similar to conventional solar cells. In fact, purely planar structures without the requirement of nanostructured $TiO_2$ have been demonstrated with efficiencies approaching those of nanostructured cells [9]. At the time of this review, however, the highest record efficiencies still are those associated with the nanostructured approach with origins in the DSSC architecture [37].

### *1.3.3 Nanostructures for Improved Optical Performance*

In order to approach or surpass the SQ limit of efficiency, all the available photons above the bandgap need to be absorbed and collected. Light management in solar cells focuses on the former, that of absorbing all the photons available. Absorption in

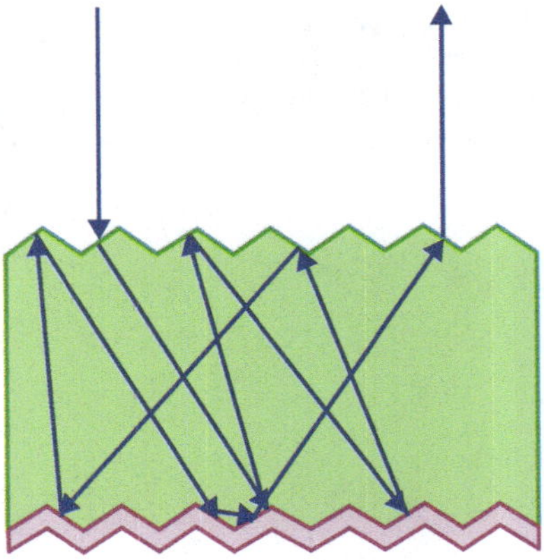

**Fig. 1.11** Illustration of the classical path of light in a general semiconductor absorber structure including texturing of the front, the back, and a back reflector

semiconductors is primarily based on the absorption coefficient, $\alpha(h\upsilon)$, which is roughly the inverse of the absorption depth in a material. It is a strong function of wavelength, starting from zero at the band edge (ideally), and for shorter wavelengths, increasing in value, so that for very short wavelengths approaching the ultraviolet, the absorption depth may be just a few tens of nanometers. The fraction of photons absorbed at a given wavelength may be written as

$$f = [1 - R(\lambda)]\left(1 - e^{-\alpha(\lambda)l}\right) \tag{1.7}$$

where $R$ is the reflection coefficient from the front surface and $l$ is the path length of photons in the semiconductor before exiting. There are two main factors to optimize: one is to minimize the reflectance loss from the front of the device, and the second is to maximize the $\alpha l$ product such that the second term is close to zero. Part of the light management strategy with respect to nanostructuring is to minimize reflectance, while another part is concerned with increasing the effective optical path length and effective absorption coefficient.

Consider the geometry shown in Fig. 1.7 of a general absorber, which in general has a backside reflector and textured front and back surfaces. What is shown is the classical trajectory of a light ray incident on the surface (and not reflected), as it passes through the absorber and out again. If the width of the absorber is $W$, for a smooth surface, with no back reflector, $l = W$, and with perfect reflection, $l = 2\ W$ (assuming perfect transmission out the front surface) (Fig. 1.11).

In the case of a material with surface and back texturing, we see that the ray is scattered randomly and may make multiple passes through the material. Using statistical arguments based on diffusive scattering from the surfaces and ray optics, the limit to which the effective path length through the material may be enhanced is limited by the so-called classical light-trapping limit [38]

$$\langle l \rangle = 4n_s^2 W \tag{1.8}$$

where $n_s$ is the index of refraction of the semiconductor. For example, in the case of Si, for red light, $n_s \approx 3.8$, which means the maximum enhancement of the path length is about 60 times.

For materials like Si which is an indirect bandgap materials, the absorption coefficient is relatively small for long-wavelength photons, requiring more than 200 microns or more of material to capture photons (not to mention long diffusion lengths to capture the photogenerated carriers), which add to the material cost as well as performance. Organic materials also suffer from poor absorption for long-wavelength photons. So much of the focus in terms of light management is on this longer-wavelength portion of the spectrum.

Nanostructured materials offer the potential to go beyond the classical light-trapping limit with feature sizes that are smaller than the characteristic wavelengths of light, and hence being in a regime of diffraction-limited optics. One interesting case is when we have periodic arrays of scatterers which coherently interact to produce *photonic bandgap* materials [39, 40]. Just as in the quantum picture of solids that the periodic potential of the crystal lattice modifies the free-electron dispersion and opens up energy gaps, likewise a periodic array of dielectric scatterings has the same effect on the optical dispersion, creating "bandgaps" in the optical spectra, creating passbands and stopbands for various ranges of frequencies. Such arrays can then be used to reflect or selectively enhance absorption in certain ranges of frequencies, allowing one to exceed the classical light-trapping limit.

Figure 1.12 shows an example of the measured reflectance from Si nanopillars fabricated using nanosphere lithography, in which silica nanoparticles are dispersed in a close-packed structure on the surface of Si and then regularly spaced nanopillars formed using reactive ion etching through the nanospheres [41]. The result is shown in the micrograph on the left side of the picture. The right side shows the measured reflectance (dashed curves) from a bare Si surface compared with two different

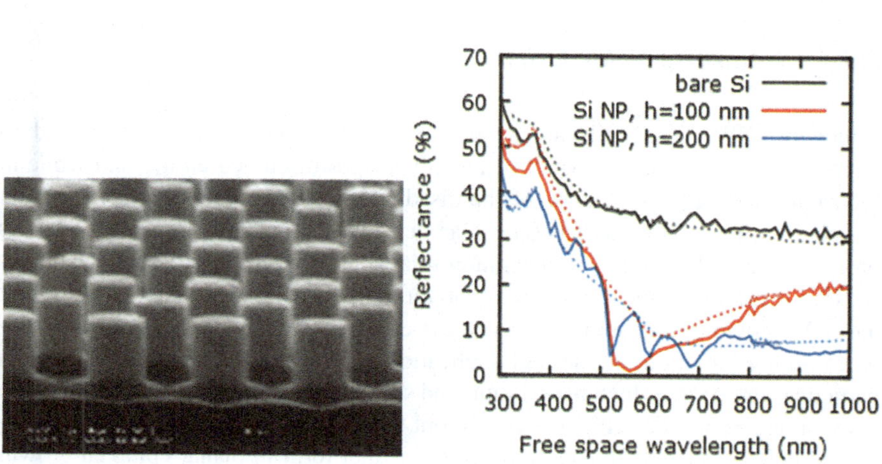


**Fig. 1.12** Left: Micrograph of nanosphere lithographically defined nanopillars. Right: FDTD simulated (solid) and measured (dashed) reflectance spectra from a regular hexagonal array of Si NPs with period, $p = 600$ nm, for cylinder heights of 100 (red) and 200 nm (blue) [34]

nanopillar heights and the numerical simulation using full-wave finite-difference time-domain (FDTD) simulation of the scattering electromagnetic waves. As can be seen, the nanopillars significantly decrease the reflectance due to light trapping and that this reflectance is well described by the full-wave electromagnetic solutions to Maxwell's equations (as opposed to ray optics). The additional features in the numerical simulation are due to the sharp features assumed in the simulated geometry as opposed to smoothing of the pillars due to the etching process. A systematic study of different nanowire arrays in terms of periodic versus random in terms of the absorption of photons has demonstrated the possibility of exceeding the classical light-trapping limit in Eq. (1.8) over a limited band of optical frequencies [42].

A plasmonic structure for solar cells consists of nanoparticles on a surface or interface, typically consisting of small metal nanoparticles (Au, Ag, Al, Cu, etc.). Here surface photonic modes (polaritons, which are quasiparticles formed by the strong interaction between an EM wave and dipoles excitations in the solid like optical phonons) are coupled with charge oscillations in the metal nanoparticles (plasmons) to form what are called surface plasmon polariton modes. In particular, the nanoparticles have strong dispersion in the vicinity of the plasma frequency of the metal electrons, and incident light can strongly couple to these resonant modes, and are strong scattering along the surface, increasing the absorption and effective optical path length. The plasmonic structure may be introduced on the top surface, bottom surface, or within the active volume of the solar cell to increase absorption. Plasmonics has been investigated in both Si and III–V solar cells, as well as organic cells, demonstrating increased light trapping in the infrared regions. For Si solar cells, for example, path length enhancements of 7–8 times have been reported [43]. Plasmonic nanoparticle arrays on GaAs cells showed increase in short-circuit current of 8% [44].

### *1.3.4 Nanowire Solar Cells*

Nanowire (NW) solar cells are a good example of nanotechnology applied to photovoltaics, illustrating several of the advantages discussed earlier and while at the same time addressing some of the disadvantages by mitigating recombination issue at surfaces. Nanowire-based solar cells have emerged in recent years as promising candidates for next-generation solar cells [45–48]. One of the advantages of NWs is the ability to tailor the bandgap through the geometry and composition of the NW, providing the ability to match the electronic and optical absorption properties during growth to the desired application, which bulk materials cannot do. In particular, due to their high aspect ratio and small cross sections, NWs can alleviate stress along their sidewall surfaces without forming detrimental lattice-mismatch-related defects, such as threading dislocations that form in planar epitaxial growth beyond a critical thickness. This property of NWs makes it possible to grow nanowire arrays on substrates with large lattice mismatch, as well as grow NW heterojunctions of highly lattice-mismatched materials, which would not be possible

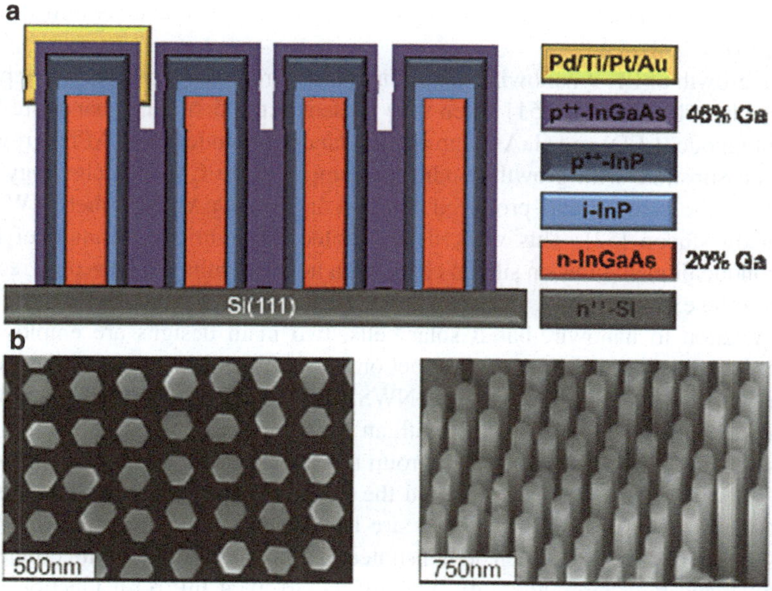


**Fig. 1.13** (**a**) Schematic of InGaAs core-shell nanowire solar cells and (**b**) SEM picture of a fabricated array [50]

in planar structures. At the same time, as discussed in the preceding section, arrays of NWs strongly modify the optical properties of the system due to the photonic bandgap materials where the NWs can be thought of as "antennas" which strongly localize optical modes and lead to greatly enhanced absorption. As such, arrays of NWs provide strong optical absorption with a fraction of the material volume required in a bulk absorber.

Until now the best performance in NW solar cells have been demonstrated in III–V compound semiconductors like GaAs, due to the large absorption coefficient and excellent transport properties of the III–Vs. However, the requirement of III–V substrates makes them very expensive, where, as discussed earlier, the main application is restricted to space applications of solar cells. On the other hand, III–V nanowire arrays can be grown on cheap substrates opening the path for novel devices. A schematic of an InGaAs NW array solar cell architecture is shown in Fig. 1.13. The particular geometry shown is based on a radial core-shell design used for other optoelectronic applications by Treu et al. [49]. Almost all NW solar cells are fabricated by growing vertical arrays of NWs patterned top down, in the particular case of Fig. 1.13, using nanoimprint lithography, which is relatively inexpensive for nanopatterning. The challenge is to make contact to the n- and p-regions of the device. One strategy is to fill in the regions between the NWs with a spin-on insulator like PDMS (polydimethylsiloxane) and then to contact the tips of the NWs with a conducting oxide like ITO. The scheme shown below contacts the $p^{++}$ outer shell InP with a thin cap of $p^{++}$ InGaAs which connects the cells in parallel, which are then contacted by a gold-alloy finger structure (the n-contact is made to the doped Si) (Fig. 1.13).

The growth of III–V nanowires on Si for optoelectronic applications was reported in 2010 by Chuang et al. [51] when they presented the first GaAs nanowire light-emitting diode (LED) and GaAs nanowire avalanche photodetector (APD) grown on a silicon substrate with growth conditions compatible to CMOS technology. One year later the same group produced the first InGaAs/GaAs core-shell NW laser grown on silicon [52]. This work demonstrated the hybrid integration of III–V semiconductor nanowires on silicon chips, with huge potential that has only recently begun to be exploited.

In relation to nanowire-based solar cells, two main designs are employed as discussed earlier: radial, core-shell junction and the axial junction devices, either p-n or p-i-n. In the case of core-shell NWS, Colombo et al. reported single GaAs nanowire p-i-n NW solar cells [53] with an efficiency of 4.5% and a good $V_{oc}$ for GaAs cells of almost 1 V. The same group more recently reported single-nanowire solar cells with the potential to exceed the Shockley-Queisser limit [54]. Besides single-nanowire SCs, nanowire arrays are the subject of intense research. Radial GaAs junction arrays have been demonstrated by Mariani et al. [55] with efficiencies of 2.54% and high reproducibility. State-of-the-art axial InP p-i-n junction arrays have reached a record 13.8% efficiency as shown by Borgström et al. [56]. Sol Voltaics and Lund University reported GaAs VLS-grown nanowire solar cells with 15.3% efficiencies which were independently verified [57]. More recently, Eindoven University reported a 17.8% InP vertical junction nanowire solar cell formed by etching and passivation [58]. As can be seen, the improvement in NW solar cell technology has rapidly evolved, from a few percent in 2009 to over 17% in 2017, not dissimilar to the rapid improvement of perovskite solar cells, although surprisingly the technology does not appear on the NREL efficiency charts (Fig. 1.5) at present.

### *1.3.5 Multiexciton Generation*

Figure 1.14 illustrates the creation of multiple electron-hole pairs for different photon energies, assuming simplistically that all the excess energy goes into the electron kinetic energy in the conduction band. As can be seen, there are different thresholds reached in energy when the photon energy in this picture is $h\nu=2E_g$, $3E_g$, $4E_g$, etc., resulting in 2EHPs, 3EHPs, etc. As we discuss below from detailed balance, increasing the quantum efficiency above 100% through multiple EHP creation allows one to exceed the SQ limit of a single-gap system.

Generation of multiple electron-hole pairs has been known in bulk materials since the 1960s in Ge and demonstrated experimentally in bulk silicon solar cells [11]. However, impact ionization or Auger generation processes have a low efficiency in bulk materials and too high a threshold energy for effective utilization of the solar spectrum due to crystal momentum conservation. Nanostructured materials have been shown experimentally to increase the efficiency of carrier multiplication

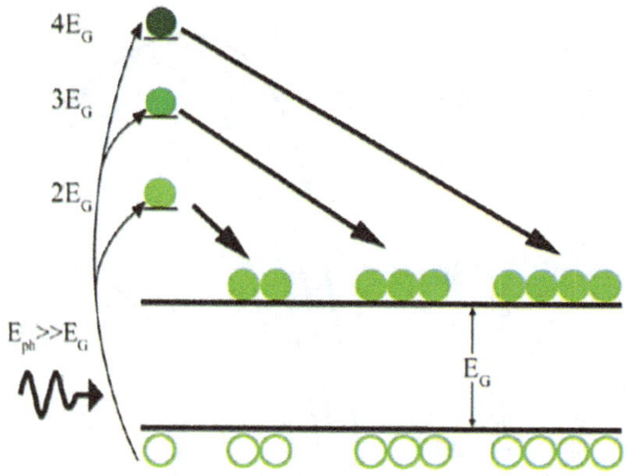


**Fig. 1.14** Illustration of the multiexciton generation process for $M = 1, 2, 3$, and 4

processes, with lower thresholds for carrier multiplication, and experimental demonstration of multiple exciton generation (MEG) in materials such as PbSe and PbS colloidal quantum dots [59, 60] with quantum efficiencies well in excess of 300%. The improved performance in nanocrystals over bulk systems is due to the relaxation of crystal momentum conservation in quantum dots, which, in bulk systems together with energy conservation, make the threshold for carrier multiplication roughly a factor of 1.5 higher than the bandgap. Due to quantum confinement, crystal momentum is no longer a good quantum number, and the threshold for carrier multiplication occurs at roughly multiples of the bandgap itself. Recent experimental evidence [61], as well as theoretical calculations [62], suggests indeed that the multi-excitation of several electron-hole pairs by single photons in quantum-dot structures occurs at ultrashort time scales, without the necessity of impact ionization. Overall, MEG generation has been shown in multiple materials, including PbSe, PbS, InAs [63], PbTe [64], Si [65], and CdSe [66].

The increase in the efficiency of a solar cell due to MEG processes may be calculated using the detailed balance approach of Eq. (1.5), by multiplying the integrand of the first integral corresponding to the incident photon flux by a quantum efficiency, $Q(E)$, representing the number of EHPs per photon generated due to impact ionization of multiexciton generation. For the case in which all the kinetic energy of the photon appears in the conduction band, we can write this mathematically as

$$Q(E) = \sum_{m=1}^{M} m\Theta(E - mE_g) \tag{1.9}$$

where $Q$ is the quantum efficiency, $m$ is the number of electron-hole pairs generated by a photon, $E_g$ is the threshold energy (which is ideally equal to the bandgap energy), $M$ is the maximum number of electron-hole pairs generated per an incident photon, and $\Theta$ represents the unit step function. Assuming a blackbody spectrum

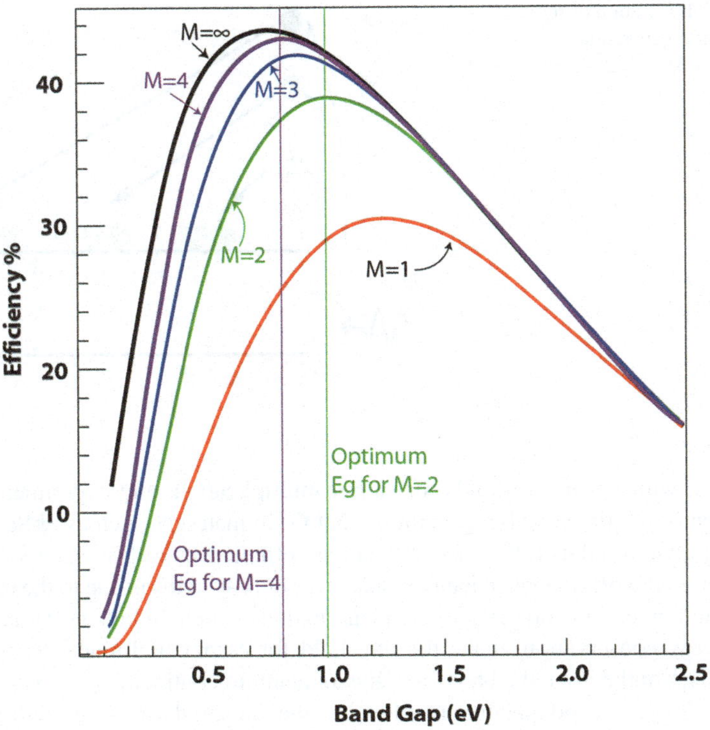


**Fig. 1.15** Calculated detailed balance efficiency as a function of bandgap for the AM0 blackbody spectrum with consideration of increasingly higher-order multiplication factors

(the result is similar of the AM1.5 spectrum), the result of detailed balance using the quantum efficiency of Eq. (1.9) is given in Fig. 1.15.

The optimum bandgap for a completely ideal MEG device is 0.76 eV [67] for $M$ going to infinity, whereas for $M$ limited to 2, the optimum bandgap is 1.05 eV, very close to that of Si. However, ideal quantum efficiency given by the step function of Eq. (1.9) assumes the existence of multiple separate, non-interacting MEG generation processes, i.e., the band structure must be ideal for generating two excitons, as well as for three excitons, and so on, and also assumes that each of these MEG processes does not interact. Further, due to the high energies involved, the photon energy is more evenly split between electrons in the conduction band and holes in the valence band, leading to a smearing out of the sharp threshold for successive MEG events.

Experimentally, the quantum yield measured using ultrafast spectroscopy shows a threshold higher than $2E_g$, with a finite slope as shown in the quantum yield data from NREL [59] shown in Fig. 1.16. As can be seen, the quantum yield in bulk Si is relatively low with a threshold voltage beyond 3.5 eV, whereas for Si nanoparticle, the threshold is much lower and the slope steeper. More recently, PbSe quantum-dot

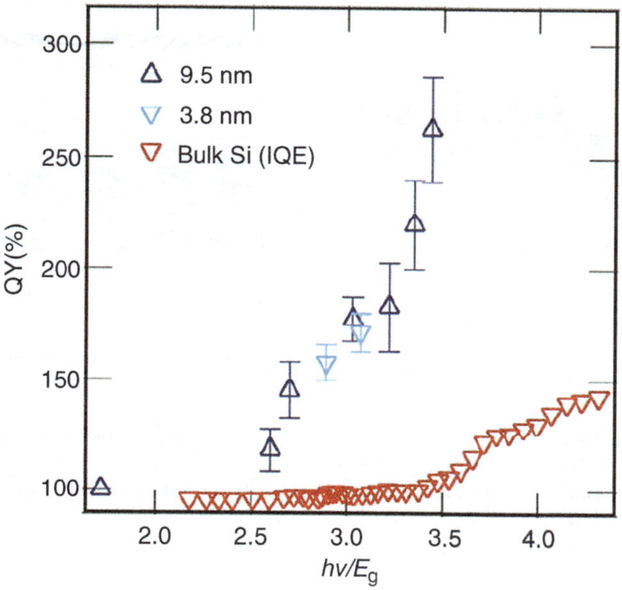


**Fig. 1.16** Measured quantum yield versus photon energy (normalized by the bandgap) for bulk Si and two different diameters of nanoparticles. (Reprinted with permission from Beard et al. [66]. Copyright 2007 American Chemical Society)

solar cells were reported in which for short wavelengths, quantum efficiencies (correct for reflection) greater than 100% were measured [68].

Multiexciton generation has been measured more recently in PbSe nanowire/nanorod structures [69], where the threshold for enhanced quantum yield was lower than that of nanoparticles of the same material. This result is quite promising, as nanowires allow transport long the axis of the nanowire, allowing efficient collection of the generated EHPs, in contrast to nanoparticles, which generally require some sort of tunneling process to extract carriers.

In terms of multiexciton generation, there is competition between the impact excitation process (assuming that it is an incoherent process) and other energy relaxation mechanisms such as electron-phonon scattering, which is responsible for thermalization, one of the two major losses discussed earlier responsible for the single-gap SQ limit. The trade-offs between phonon scattering and impact ionization in narrow nanowires (1–5 nm) have been investigated theoretically using ensemble Monte Carlo simulation, a particle-based technique for simulating the nonequilibrium dynamics of photoexcited electrons and holes [70]. Figure 1.17 shows the calculated scattering rates based on the electronic states in the nanowire from an atomistic sp3d5s* tight-binding representation. What are shown are the bulk versus nanowire rates for deformation potential scattering (energy averaged) and polar optical phonon scattering, the two major lattice relaxation processes in III–V materials. As can be seen, the nanowire rates at low energy deviate strongly due to the highly 1D nature of the electronic states, whereas at high energies, the nanowire

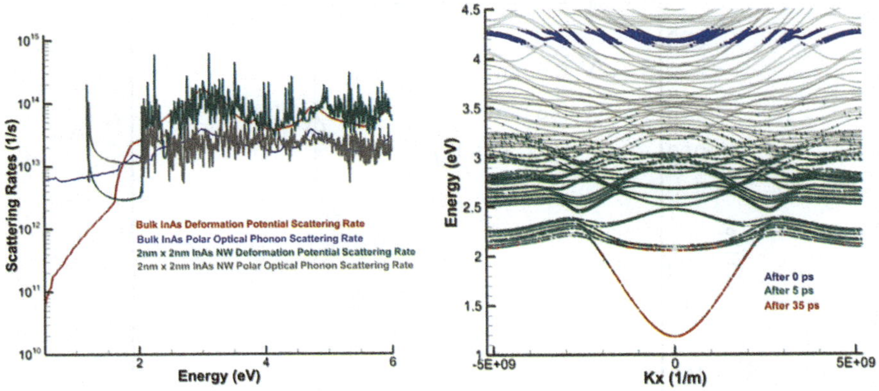


**Fig. 1.17** Calculated scattering rates in 2 nm InAs NWs compared to the bulk scattering rates for deformation potential scattering and polar optical scattering (left panel). The simulated relaxation with the NW band structure for different snapshots in time starting with a high-energy nonequilibrium distribution [70]

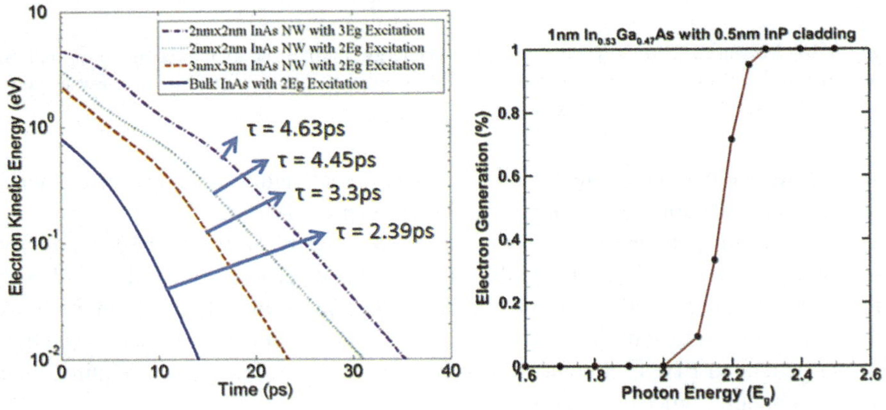


**Fig. 1.18** Average energy versus time for a $2E_g$ excitation of carriers in the conduction band of different size nanowires versus bulk (left panel). Corresponding simulated quantum yield versus excitation energy

bands merge together to approximate the bulk density of states, and hence the same corresponding scattering rates. The right panel of Fig. 1.17 shows snapshots at various times of the time evolution of a nonequilibrium carrier distribution due to high-energy photoexcitation, which shows that as carriers relax, a bottleneck occurs in going to the ground state due to the energy separation and reduced scattering rates, leading to longer energy relaxation.

In Fig. 1.18, the average electron kinetic energy of photoexcited carriers in the conduction band of the InAs NW as a function of time for different nanowire dimensions compared with the bulk rate is shown on the right side. As can be seen, the energy loss rate is substantially reduced in the NW system, almost double that of bulk InAs, and increases for increasingly smaller NWs. The corresponding electron generation due to impact excitation is shown on the rate, showing that at nearly $2E_g$, there is a strong threshold for carrier generation, which indicates that NWs should be good candidates for MEG-based solar cells.

### *1.3.6 Hot-Carrier Solar Cells*

Ross and Nozik proposed the concept of hot-carrier solar cells [71] more than 25 years ago as a means to circumvent the limitations imposed by the Shockley-Queisser limit in terms of both the loss of excess kinetic energy and the loss of sub-bandgap photons. Figure 1.19 shows a schematic of the basic concept. The ideal absorber represents a material with a bandgap, $E_G \geq 0$, across which electron-hole pairs are excited by photons with energies greater than $E_G$. In the absorber, the relaxation of excess kinetic energy to the environment (i.e., the lattice) is suppressed, while the carriers themselves still interact strongly to establish a thermalized distribution, such that the electrons (and holes) are characterized by an effective temperature, $T_H$, much greater than the lattice temperature, $T_L$. This carrier temperature can be so large as to reverse the net chemical potential difference, $\mu_{ch}$, between electrons and holes, and typically must be on the order of several thousand degrees for efficient operation.

Energy-selective contacts are made to the absorber on the left and right, where the left contact extracts hot electrons in a narrow range of energies above the conduction band edge as shown, while the contact on the right extracts holes (injects electrons) at a specific energy range in the valence band. In this scheme, the electrons and hole are extracted from the system before they have time to relax their excess energy, hence utilizing the total energy of the photon. Under the assumption of no energy loss, the maximum efficiency occurs for vanishingly small bandgaps, hence capturing photons over the entire solar spectrum. In this limit, the theoretical detailed balance conversion efficiency approaches the maximum thermodynamic conversion efficiency of 85.4% [72]. Later, Würfel and coworkers considered the effect of

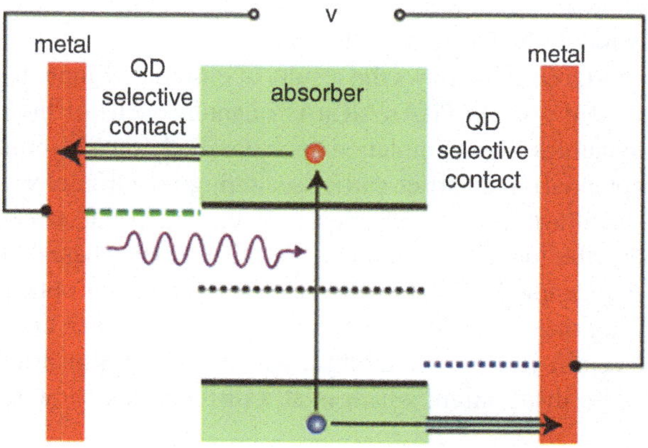


**Fig. 1.19** Schematic of a hot-carrier solar cell consisting of an ideal absorber with energy-selective contacts

impact ionization and secondary carrier generation on the ultimate efficiency of this concept [73, 74].

There are many practical limitations to implementing this very ideal structure. One difficulty is realizing energy-selective contacts. Würfel pointed out [53] that it is necessary to spatially separate the absorber material for the cold metallic contacts themselves, which may serve as an energy loss mechanism to the carriers in the absorber layer. There it was suggested that a large bandgap material such as GaN serves as a spacer or "membrane" separating the absorber from the contacts. Other proposals for energy-selective contacts include using nanostructured resonant tunneling contacts from double-barrier heterostructures, defects, or artificial quantum dots [75].

The main challenge in the technology is to realize an ideal absorber in which the excess kinetic energy of the photoexcited carriers is not lost to the environment. There have been various proposals for reducing the carrier cooling rate. Due to the reduced dimensionality and therefore reduced density of final states in nanostructured systems, the energy loss rate due to phonons may be reduced, which has been observed experimentally [76]. In particular, in nanostructured systems such as quantum wells, quantum wires, or quantum dots, where intersubband spacing between levels is less than the optical phonon energy, then the optical emission rate may be suppressed due to the so-called "phonon bottleneck" effect, since there is no final states for the electron. However, even in such systems, the reduced phonon emission rate is still too fast for sufficient carrier heating, even under high solar concentration. As we saw in the previous section on energy relaxation in NWs, the energy loss rate is reduced by a factor of two compared to bulk, although this rate is still too fast of establishing a steady-state hot-carrier distribution in the absorber.

If, however, the energy is retained in the coupled electron-phonon system, then the energy may be recycled through hot-phonon reabsorption. Nonequilibrium hot-phonon effects during ultrafast photoexcitation have been well studied for many years. Time-resolved Raman scattering has been used, for example, to characterize the optical phonon decay after photoexcitation for a variety of III–V compound bulk and quantum well materials [77–80]. Ensemble Monte (EMC) simulation has previously been used to theoretically model ultrafast carrier relaxation and hot-phonons effects in quantum well and bulk materials [81, 82], where hot phonons have been shown to significantly reduce the rate of carrier cooling compared to the bare energy loss rate.

Figure 1.20 shows the results of ensemble Monte Carlo simulation of a quasi-2D system (10 nm GaAs/AlGaAs quantum well) of the early-stage carrier relaxation dynamics. The simulation includes both optical phonon (polar and deformation potential) and carrier-carrier scattering (electron-electron, hole-hole, electron-hole) and illustrates several effects. In the left panel, we see the early athermal carrier distribution that is more or less Gaussian shaped around the injection energy. A secondary peak already appears, which is an optical phonon replica of the main peak due to the short emission time. For longer times approaching a picosecond, electron-electron scattering drives the athermal distribution toward a heated Fermi-Dirac distribution, which is at a different temperature than the lattice, and over a

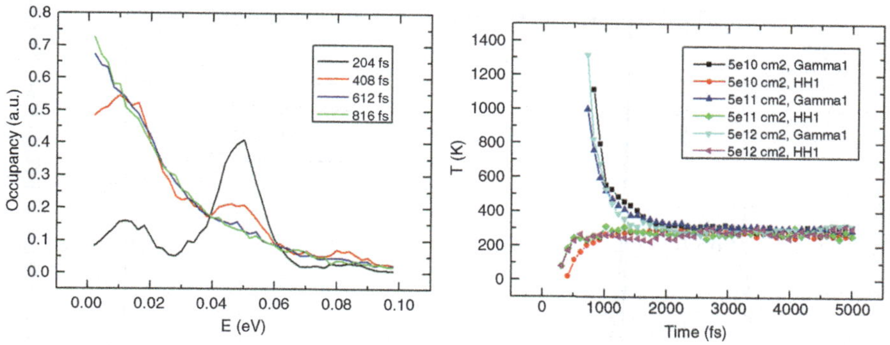


**Fig. 1.20** Left panel: Ensemble Monte Carlo simulation of the ultrafast electron dynamics in the conduction band of a 10 nm GaAs quantum well at 300 K for carriers injected at 50 meV above the band edge, with an injected carrier density of $5\times10^{11}/\text{cm}^2$. Right panel: Electron and hole temperatures versus time after ultrafast carrier excitation for different injected carrier densities

much longer time scale (see, e.g., Fig. 1.18). The right panel shows the evolution of the electron and hole temperatures (taken from the average energy), where initially the electron and hole average temperatures are very different. However, over a period of 2 ps, the two temperatures reach a common temperature due to electron-hole scattering, which exchanges energy between the two systems.

Basically the main energy relaxation channel for electrons is through optical phonons, which lose energy through optical phonon emission in quanta of the optical phonon energy. However, due to the small group velocity of optical phonons, they do not leave the excitation volume; rather they must decay into acoustic phonons through a three-phonon anharmonic scattering process, and it is the acoustic phonons which propagate energy away from the active region of the device. Hence electrons and holes may reabsorb the excess phonons, and so the excess kinetic energy of the photoexcited EHPs remains in the system until the optical phonons decay to acoustic modes. It has been argued by the UNSW group that nonequilibrium "hot" phonons may play a critical role in reducing carrier energy loss and maintaining energy within the absorber [83]. Typical optical phonon decay times range from 1 to 10 ps, much longer than the electron-optical phonon emission rate (which is subpicosecond in scale). Engineering materials as absorbers with long phonon decays, particularly nanoengineered structures, are currently being investigated [56].

Figure 1.21 shows the simulated effect of phonon lifetime on carrier relaxation using EMC simulation, similar to earlier work on this topic [60, 84]. Here a 2 eV laser pulse exciting a 10 nm GaAs/AlAs QW is simulated, which peaks at 1 ps into the simulation, and is 200 fs wide. Optical absorption is modeled by creating electron-hole pairs corresponding to photons with a given frequency and momentum. Figure 1.21 plots the carrier temperature as a function of time for various assumed phonon lifetimes ranging from 0 (i.e., no hot phonons) to 100 ps.

As can be seen in the simulated results of Fig. 1.21, without hot phonons, the electrons cool rapidly and reach the lattice temperature within 5–10 ps. In contrast, with hot phonons, after the initial pulse, when a nonequilibrium distribution of hot

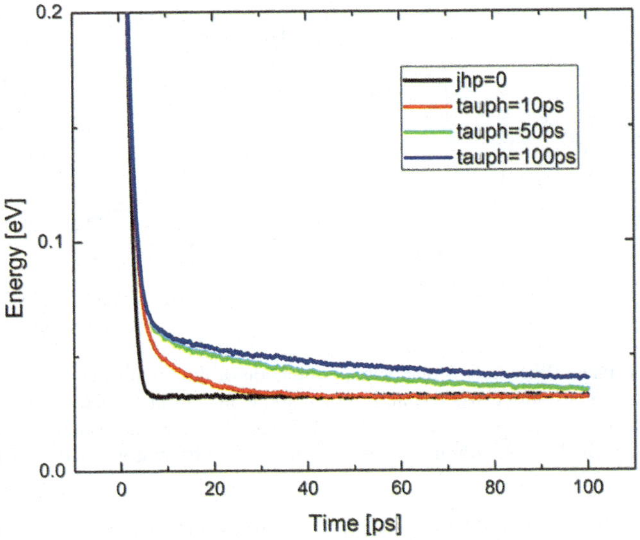


**Fig. 1.21** Simulated electron temperature versus time for various assumed phonon lifetimes in a 10 nm GaAs/AlAs QW following a 2 eV, 200 fs wide optical pulse. The injected carrier density is $5\times10^{11}/\ \text{cm}^2$ in all cases. The lattice temperature is 5 K

phonons establishes itself, the decay slows and becomes non-exponential. As expected, as the phonon decay time becomes very large, the energy loss rate transitions to one limited by the electron-phonon scattering time, to one determined by the phonon-phonon anharmonic decay time. Hence, finding absorber materials with long phonon lifetimes is a possible approach to realizing hot-carrier solar cell performance.

## 1.4 Summary

Here we have discussed the state of the art with respect to photovoltaic device technology and how nanotechnology is playing an increasing role in improving existing devices, as well as new device architectures seeking to improve efficiency while lower cost. We discussed how, for example, nanostructures are playing an increasing role in improving light management in solar cells to improve light collection and allow thinner materials to be used reducing cost. Nanostructured materials play a central role in device architectures such as the dye-sensitized solar cell, which in turn became the basis for perovskite solar cells, which have rapidly overtaken thin-film technology in terms of efficiency and approach that of crystalline Si solar cells. Nanowire solar cells have also shown tremendous improvement in recent years, with efficiencies over 15%. Research continues on realizing advanced concept solar cell structures such as multiexciton generation and hot-carrier solar cells, and recent results show continued improvement in the design and architectures of such systems.

## References


1. Photovoltaics Report (PDF). Fraunhofer ISE. 12 July 2017
2. Yoshikaw K, Kawasaki H, Yoshida W, Irie T, Konishi K, Nakano K, Uto T, Adachi D, Kanematsu M, Uzu H, Yamamoto K (2017) Silicon heterojunction solar cell with interdigitated back contacts for a photoconversion efficiency over 26%. Nat Energy 2:1–8
3. https://www.altadevices.com/
4. "First solar builds the highest efficiency thin film PV cell on record". firstsolar.com
5. Dennler G, Scharber MC, Brabec CJ (2009) Polymer-fullerene bulk-heterojunction solar cells. Adv Mater 21:1323–1338
6. Timmreck R, Meyer T, Gilot J, Seifert H, Nueller T, Furlan A, Wienk MM, Wynands D, Hohl-Ebinger J, Warta W, Janssen RAJ, Riede M, Leo K (2015) Characterizaion of tandem organic solar cells. Nat Photonics 9:478–479
7. Grätzel M (2014) The light and shade of perovskite solar cells. Nat Mater 13:838–842
8. Kojima A, Teshima K, Shirai Y, Miyasaka T (2009) Organometal halide perovskites as visible-light sensitizers for photovoltaic cells. J Am Chem Soc 131:6050–6051
9. Liu M, Johnston MB, Snaith HJ (2013) Efficient planar heterojunction perovskite solar cells by vapour deposition. Nature 501:395–398
10. Shockley W, Queisser HJ (1961) Detailed balance limit of efficiency of p-n junction solar cells. J Appl Phys 32:510–519
11. de Vos A (1992) Endoreversible thermodynamics of solar energy conversion. Oxford University Press, Oxford
12. Kolodinski S, Werner JH, Wittchen T, Queisser HJ (1993) Quantum efficiencies exceeding unity due to impact ionization in silicon solar cells. Appl Phys Lett 63:2405
13. Schaller R, Klimov V (2004) High efficiency carrier multiplication in pbse nanocrystals: implications for solar energy conversion. Phys Rev Lett 92:186601
14. Ross RT, Nozik AJ (1982) Efficiency of hot-carrier solar energy converters. J Appl Phys 53:3813
15. Würfel P (1997) Solar energy conversion with hot electrons from impact ionization. Sol Energy Mater Sol Cells 46:43
16. Würfel P, Brown AS, Humphrey TE, Green MA (2005) Particle conservation in the hot-carrier solar cell. Prog Photovolt Res Appl 13:277
17. Cotal H, Fetzer C, Boisvert J, Kinsey G, King R, Hebert P, Yoon H, Karam N (2009) III–V multijunction solar cells for concentrating photovoltaics. Energy Environ Sci 2:174–192
18. Press Release, Fraunhofer Institute for Solar Energy Systems, 1 December 2014 (accessed at http://www.ise.fraunhofer.de/en/ press-and-media/press-releases/press-releases-2014/new-world-record-for-solar-cell -efficiency-at-46-percent on 7 December 2014). Soitec record
19. Zhang C, Kim Y, Faleev NN, Honsberg CB (2017) Improvement of GaP crystal quality and silicon bulk lifetime in GaP/Si heteroepitaxy. J Cryst Growth 475(1):83–87
20. Weyers M, Sato M, Ando H (1992) Red shift of photoluminescence and absorption in dilute GaAsN alloy layers. Jpn J Appl Phys 31(pt. 2, 7A):L853–L955
21. Luque A, Martí A (1997) Increasing the efficiency of ideal solar cells by photon induced transitions at intermediate levels. Phys Rev Lett 78:5014–5017
22. Binnig G, Rohrer H, Gerber C, Weibel E (1982) Surface studies by scanning tunneling microscopy. Phys Rev Lett 49:57–60
23. Cho AY, Arthur JR (1975) Molecular beam epitaxy. Prog Solid State Chem 10:157–191
24. Lu W, Lieber CM (2006) Semiconductor nanowires. J Phys D Appl Phys 39:R387
25. Samuelson L (2003) Self-forming nanoscale devices. Mater Today 6:22–31
26. Björk MT, Ohlsson BJ, Sass T, Persson AI, Thelander C, Magnusson MH, Deppert K, Wallenberg LR, Samuelson L (2002) One-dimensional steeplechase for electrons realized. Nano Lett 2:87–89
27. Björk MT, Ohlsson BJ, Thelander C, Persson AI, Deppert K, Wallenberg LR, Samuelson L (2002) Nanowire resonant tunneling diodes. Appl Phys Lett 81:4458–4460

28. Thelander C, Martensson T, Björk MT, Ohlsson BJ, Larsson MW, Wallenberg LR, Samuelson L (2003) Single-electron transistors in heterostructure nanowires. Appl Phys Lett 83:2052–2054
29. Fuhrer A, Fasth C, Samuelson L (2007) Single electron pumping in InAs nanowire double quantum dots. Appl Phys Lett 91:052109
30. Fuhrer A, Froberg LE, Pedersen JN, Larsson MW, Wacker A, Pistol ME, Samuelson L (2007) Few electron double quantum dots in InAs/InP nanowire heterostructures. Nano Lett 7:243–246
31. Björk MT, Fuhrer A, Hansen AE, Larsson MW, Fröberg LE, Samuelson L (2005) Tunable effective g factor in InAs nanowire quantum dots. Phys Rev B 72:201307
32. Chandra N, Tracy CJ, Cho J-H, Picraux ST, Hathwar R, Goodnick SM (2015) Vertically grown Ge nanowire Schottky diodes on Si and Ge substrates. J Appl Phys 118:024301–024307
33. Alivisatos AP (1996) Perspectives on the physical chemistry of semiconductor nanocrystals. J Phys Chem 100:13226–13239
34. Bimberg D, Grundmann M, Ledentsov NN (1999) Quantum dot heterostructures. Wiley, Chichester
35. Ban K-Y, Bremner SP, Liu G, Dahal SN, Dippo PC, Norman AG, Honsberg CB (2010) Use of a GaAsSb buffer layer for the formation of small, uniform, and dense InAs quantum dots. Appl Phys Lett 96:183101
36. O'Regan B, Grätzel M (1991) A low-cost, high-efficiency solar cell based on dye-sensitized colloidal $TiO_2$ films. Nature 353:737–740
37. Seok SI, Grätzel M, Park N-G (2018) Methodologies toward Highly Efficient Perovskite Solar Cells. Small 14:1704177
38. Yablonovitch E (1982) Statistical ray optics. J Opt Soc Am A 72:899–907
39. John S (1987) Strong localization of photons in certain disordered dielectric superlattices. Phys Rev Lett 58(23):2486–2489
40. Yablonovitch E (1987) Inhibited spontaneous emission in solid-state physics and electronics. Phys Rev Lett 58(20):2059–2062
41. Vulic N, Choi J-Y, Honsberg CB, Goodnick SM (2015) Silica nanosphere lithography defined light trapping structures for ultra-thin Si photovoltaics. MRS Proc 1770:31–36
42. Kelzenberg MD, Boettcher SW, Petykiewicz JA, Turner-Evans DB, Putnam MC, Warren EL, Spurgeon JM, Briggs RM, Lewis NS, Atwater HA (2010) Enhanced absorption and carrier collection in Si wire arrays for photovoltaic applications. Nat Mater Lett 9:239–244
43. Pillai S, Catchpole KR, Trupke T, Green MA (2007) Surface plasmon enhanced silicon solar cells. J Appl Phys 101(9):093105
44. Nakayama K, Tanabe K, Atwater HA (2008) Plasmonic nanoparticle enhanced light absorption in GaAs solar cells. Appl Phys Lett 93(12):121904
45. Tian B et al (2007) Coaxial silicon nanowires nanoelectronic power sources. Nature 449:889
46. Garnett EC, Peidong Y (2008) Silicon nanowire radial p-n junctions solar cells. J Am Chem Soc 130(29):9224–9225
47. Tang J, Huo Z, Brittman S, Gao H, Yang P (2011) Solution-processed core-shell nanowires for efficient photovoltaic cells. Nat Nanotechnol 6(9):568–572
48. Kempa TJ, Kim SK, Day RW, Park HG, Nocera DC, Lieber CM (2013) Facet-selective growth on nanowires yields multi-component nanostructures and photonic devices. J Am Chem Soc 135(49):18354–18357
49. Treu J, Stettner T, Watzinger M, Morkötter S, Döblinger M, Matich S, Saller K, Bichler M, Abstreiter G, Finley JJ, Stangle J, Koblmüller G (2015) Lattice-matched InGaAs-InAlAs core-shell nanowires with improved luminescence and photoresponse properties. Nano Lett 15 (5):3533–3540
50. Popescu B, Popescu D, Luppina P, Julian T, Koblmüller G, Lugli P, Goodnick S (2015) Modeling and simulation of InGaAs nanowire solar cells. In Proceedings of the IEEE International Conference on Nanotechnology, Rome, Italy, pp 728–231
51. Chuang C, Sedgwick FG, Chen R, Ko WS, Moewe M, Ng W, Tran T-TD, Chang-Hasnain C (2010) GaAs-based nanoneedle light emitting diode and avalanche photodiode monolithically integrated on a silicon substrate. Nano Lett 11(2):385–390

52. Chen R, Tran T-TD, Ng KW, Ko WS, Chuang LC, Sedgwick FG, Chang-Hasnain C (2011) Nanolasers grown on silicon. Nat Photonics 5(3):170–175
53. Colombo C, Heiss M, Gratzel M, Fontcuberta I Morral A (2009) Gallium arsenide pin radial structures for photovoltaic applications. Appl Phys Lett 94(17):173108
54. Krogstrup P et al (2013) Single-nanowire solar cells beyond the Shockley-Queisser limit. Nat Photonics 7(4):206–310
55. Mariani G et al (2011) Patterned radial GaAs nanopillar solar cells. Nano Lett 11(6):2490–2494
56. Wallentin J, Anttu N, Asoli D, Huffman M, Magnusson IAMH, Siefer G, Fuss-Kailuweit P, Dimroth F, Witzigmann B, Xu HQ, Samuelson L, Deppert K, Borgström MT (2013) InP nanowire array solar cells achieving 13.8% efficiency by exceeding the ray optics limit. Science 339(6123):1057–1060
57. Åberg I, Vescovi G, Asoli D, Naseem U, Gilboy JP, Sundvall C, Dahlgren A, Svensson KE, Anttu N, Björk MT, Samuelson L (2016) A GaAs nanowire array solar cell with 15.3% efficiency at 1 sun. IEEE J Photovolt 6(1):185–190
58. van Dam D, van Hoof NJJ, Cui Y, van Veldhovern PJ, Bakkers EPAM, Rivas JG, Haverkort JEM (2016) High-efficiency nanowire solar cells with omnidirectionally enhanced absorption due to self aligned indium−tin−oxide mie scatterers. ACS Nano 10:11414–11419
59. Schaller RD, Klimov VI (2004) High efficiency carrier multiplication in PbSe nanocrystals: implications for solar energy conversion. Phys Rev Lett 92(18):186601
60. Ellingson RJ, Beard MC, Johnson JC, Yu P, Micic OI, Nozik AJ, Shabaev A, Efros AL (2005) Highly efficient multiple exciton generation in colloidal PbSe and PbS quantum dots. Nano Lett 5(5):865–871
61. Nozick AJ (2005) Exciton multiplication and relaxation dynamics in quantum dots: applications to ultrahigh-efficiency solar photon conversion. Inorganic Chem 44:6893
62. Shabaev A, Efros AL, Nozik AJ (2006) Multiexciton generation by a single photon in nanocrystals. Nano Lett 6:8
63. Schaller RD, Pietryga JM, Klimov VI (2007) Carrier multiplication in InAs nanocrystal quantum dots with an onset defined by the energy conservation limit. Nano Lett 7 (11):3469–3476
64. Murphy JE, Beard MC, Norman AG, Phillip S, Johnson JCA, Pingrong M, Olga IY, Ellingson RJ, Nozik AJ (2006) PbTe colloidal nanocrystals: synthesis, characterization, and multiple exciton generation. J Am Chem Soc 128(10):3241–3247
65. Werner JH, Kolodinski S, Queisser HJ (1994) Novel optimization principles and efficiency limits for semiconductor solar cells. Phys Rev Lett 72(24):3851–3854
66. Beard MC, Knutsen KP, Yu P, Luther JM, Song Q, Metzger WK, Ellingson RJ, Nozik AJ (2007) Multiple exciton generation in colloidal silicon nanocrystals. Nano Lett 7(8):2506–2512
67. Vos A d, Desoete B (1998) On the ideal performance of solar cells with larger-than-unity quantum efficiency. Sol Energy Mater Sol Cells 51(3):413–424
68. Semonin OE, Luther JM, Choi S, Chen H-Y, Gao J, Nozik AJ, Beard MC (2012) Peak external photocurrent quantum efficiency exceeding 100% via MEG in a quantum dot solar cell. Science 334:1530–1533
69. Cunningham PD, Boercker JE, Foos EE, Lumb MP, Smith AR, Tischler JG, Melinger JS (2011) Enhanced multiple exciton generation in quasi-one-dimensional semiconductors. Nano Lett 11 (8):3476–3481
70. Hathwar R, Saraniti M, Goodnick SM (2015) Energy relaxation and non-linear transport in InAs nanowires. J Phys Conf Ser 647:012039
71. Ross RT, Nozik AJ (1982) Efficiency of hot-carrier solar energy converters. J Appl Phys 53:3813–3818
72. Landsberg PT, Tonge G (1980) Thermodynamic energy conversion efficiencies. J Appl Phys 51:R1
73. Würfel P (1997) Solar energy conversion with hot electrons from impact ionization. Sol Energy Mater Sol Cells 46:43–52

74. Würfel P, Brown AS, Humphrey TE, Green MA (2005) Particle conservation in the hot-carrier solar cell. Prog Photovolt Res Appl 13:277
75. Conibeer G, Green MA, Corkish R, Cho Y, Chob E, Jiang C, Fangsuwannarak T, Pink E, Huang Y, Puzzer T, Trupke T, Richards B, Shalav A, Lind K (2006) Silicon nanostructures for third generation photovoltaic solar cells. Thin Solid Films 511–512:654
76. Pelouch WS, Ellingson RJ, Powers PE, Tang CL, Szmyd DM, Nozik AJ (1992) Comparison of hot-carrier relaxation in quantum wells and bulk GaAs at high carrier densities. Phys Rev B 45:1450–1453
77. Tsen KS, Wald KR, Ruf T, Yu PY, Morkoc H (1991) Electron optical phonon interactions in ultrathin GaAs AlAs multiple quantum well structures. Phys Rev Lett 67:2557–2560
78. Tsen KT, Joshi RP, Ferry DK, Botcharev A, Sverdlov B, Salvador A, Morkoc H (1996) Non-equilibrium electron distributions and phonon dynamics in wurtzite GaN. Appl Phys Lett 68:2990–2992
79. Tsen KT, Kiang JG, Ferry DK, Morkoc H (2006) Subpicosecond time-resolved Raman studies of LO phonons in GaN: dependence on photoexcited carrier density. Appl Phys Lett 89:112111
80. Tsen KT, Kiang JG, Ferry DK, Lu H, Schaff WJ, Lin H-W, Gwo S (2007) Direct measurements of the lifetimes of longitudinal optical phonon modes and their dynamics in InN. *Appl Phys Lett* 90:152107-1-3
81. Goodnick SM, Lugli P (1992) Hot carrier relaxation in quasi-2D systems. In: Shah J (ed) Hot carriers in semiconductor microstructures: physics and applications. Academic Press, New York, pp 191–234
82. Dür M, Goodnick SM, Lugli P (1996) Monte Carlo simulation of intersubband relaxation in wide, uniformly doped GaAs/AlxGa1-xAs quantum wells. *Phys Rev B* 54:17794
83. Conibeer G, Patterson R, Huang L, Guillemoles J-F, König D, Shrestha S, Green MA (2010) Modelling of hot carrier solar cell absorbers. Sol Energy Mater Sol Cells 94:1516–1521
84. Goodnick SM, Honsberg C (2012) Modeling carrier relaxation in hot carrier solar cells. *Proc SPIE* 8256:82560W. https://doi.org/10.1117/12.910858

# Chapter 2
# The FinFET: A Tutorial

**Charles Dančak**

**Abstract** Ever since Intel launched its successful 22-nm *Ivy Bridge* CPU chip, establishing nonplanar finFET technology as a viable means of extending Moore's law, variations of the basic finFET or the nanowire transistor have been introduced into nanoelectronics research and manufacturing efforts at an unprecedented rate.

All members of the finFET family of devices—whether fabricated on a bulk or SOI substrate, or structured as a double- or triple-gate or a nanowire transistor—share the same fundamental operating mechanism: the electric *field effect*. Drain current is under the control of a gate voltage that modulates the conductivity of the underlying fin or nanowire. This tutorial derives qualitative *I-V* characteristics for such devices, using intuitive assumptions and principles like Ohm's law and Poisson's equation.

It then focuses on the geometric characteristics of nonplanar devices, examining the layout of a typical CMOS standard cell comprising both *n*- and *p*-type finFETs. The use of a *local interconnect layer* to connect adjacent fins is detailed. A section is devoted to explaining why nonplanar devices exhibit *higher immunity* to short-channel effects (SCEs). To quantify this explanation, the natural screening length parameter $\lambda$ is introduced and then computed for several common device types.

The tutorial cites a dozen intuitive *rules of thumb* that engineers and scientists may find useful in evaluating finFET design issues and device trade-offs. During the discussion, key electrical and physical finFET properties are related to their corresponding BSIM-CMG SPICE parameters, including `GEOMOD` and `NFIN`.

## 2.1 Beyond Planar Technology

One of the most notable hand-drawn graphs in the world of technology was the logarithmic plot tacked to a wall of Gordon Moore's office at Fairchild in Silicon Valley [1]. It revealed a trend: MOS integrated circuits doubled in complexity each year. A chemical engineer by training, Moore realized that no real physical barriers

---

C. Dančak (✉)
VLSI Engineering, UCSC Silicon Valley Extension, Santa Clara, CA, USA
e-mail: cdancak@ucsc.edu

 
S. M. Goodnick et al. (eds.), *Semiconductor Nanotechnology*, Nanostructure Science and Technology, https://doi.org/10.1007/978-3-319-91896-9_2

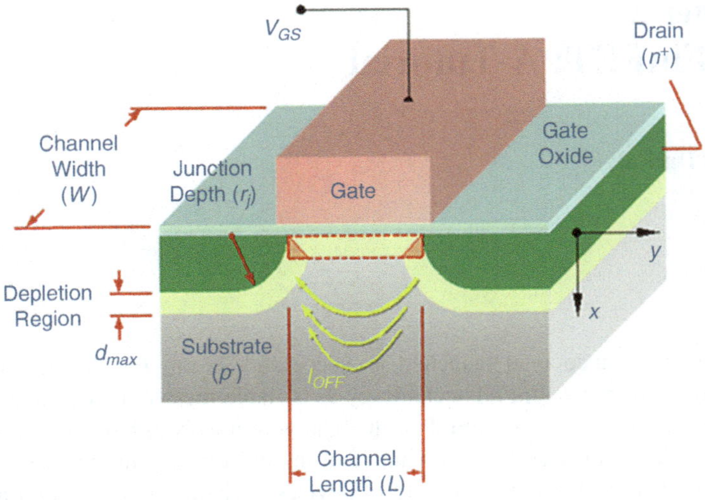


**Fig. 2.1** Short-channel *n*-type MOSFET (side view)

stood in the way of cramming more and more transistors onto a single die—unlike the thermodynamic equilibria conditions that can limit yield in chemical reactions. Achieving greater density and higher speed was just a matter of finer patterning [2].

Since 1965, Moore's law has held for nearly five decades. Chip companies like Intel have progressed from the first 4004 microprocessor, with just 2,300 MOS transistors, to processors like the Xeon, with 7 billion on one die—just by scaling down device length, width, and other process parameters by ~√2 each year or two.

What has complicated the scaling, and threatened to upset Moore's law, is a set of unwanted side effects, collectively known as short-channel effects (SCEs). SCEs arise from shrinking the channel length and other critical parameters. In 1965, the shortest commercial MOSFET had a channel length *L* of about 1 mil, or 25 μm. By 2011, Intel was planning to announce its *Ivy Bridge* CPU, with a length of 22 nm—$10^3\times$ shorter. To circumvent the SCEs, they chose to go beyond planar technology.

Figure 2.1 is a conceptual view of a traditional short-channel *n*MOS transistor. (Advanced aspects, like LDD extensions or halo doping, are omitted for simplicity.) While the channel length *L* has been scaled down aggressively, the junction depth $r_j$ of the source or drain has not scaled down as readily. One reason is the higher *sheet resistance* of shallow source-drain islands, limiting the drive current. Another is the inevitable *diffusion* of implanted dopant atoms during the fabrication process.

As a result, the short-channel transistor is no longer very planar. This is evident in Fig. 2.1. The depth of the source-drain islands ($r_j$) is now comparable to the length *L*. Since these *n*-type islands form *p*-*n* junctions with the *p*-type substrate, they are surrounded by *depletion regions* of significant depth (*d*). The shorter the device, the more these depletion regions tend to encroach into the channel area underneath the gate. (For simplicity, the figure shows depletion regions of *uniform* depth, which assumes the source-drain voltage $V_{DS}$ is small compared to VDD, and the transistor is operating in the linear region. SCEs worsen at higher drain bias.)

For older, long-channel transistors, the gate electrode effectively controlled the buildup of depletion charge (dashed rectangular outline in the figure). For shorter channels, with more encroachment, the gate lost control of some of this charge (shaded triangles). This shaded fraction of depletion charge needs *no gate voltage* to build up—it originates from the *p-n* junctions at the source and drain terminals.

What in 1965 was a rectangular volume of depletion charge under gate control (equal to $-qN_AWLd_{max}$, where $N_A$ is the substrate doping) has in recent years become *trapezoidal*. As $L$ shrinks, the shaded portions of the rectangle grow more significant. The undesirable result is that the transistor turns on at a *lower* gate voltage $V_{GS}$. The threshold voltage $V_{Tn}$ is no longer determined only by processing parameters like $N_A$ but instead decreases—*rolls off*—with shorter device length $L$.

Threshold voltage roll-off is one of the worst SCEs, in the sense that it directly impedes Moore's law. Further scaling down of $L$ can only be done at the expense of process enhancements that keep threshold voltages from rolling off excessively.

Figure 2.1 indicates another SCE. The yellow arrows superimposed on the figure occur when $V_{GS}$ is just *below* threshold. Ideally, the transistor should turn off. In reality, there is a small drain current. It is due largely to electrons *diffusing* from the electron-rich source into the electron-poor *p*-type substrate and reaching the drain.

This subthreshold current will increase as $L$ shrinks, since the channel is shorter compared to carrier diffusion length. The result is a parasitic current flow $I_{OFF}$ which contributes to the chip's static leakage power consumption in *standby* mode.

Consequently, by the 22-nm node, the traditional planar MOSFET had reached its limits. Scaling down $L$ led to too much leakage, for too little performance boost. The most notorious of the SCEs impacting its operation are summarized below [3]:

- **Threshold voltage roll-off:** The threshold voltage, at which a transistor turns on, is *no longer independent* of gate length but rolls off with decreasing $L$.
- **Subthreshold conduction:** Just below threshold, the transistor *never turns off* entirely. A minuscule current flows, by the diffusion of carriers across the short channel. Though exponentially small, this subthreshold current—multiplied by billions of on-chip devices—leads to unacceptable levels of leakage power.
- **Channel-length modulation:** An early warning sign of SCEs, channel-length modulation (CLM) results in *I-V* curves that *fail to saturate* above pinch-off. The channel current $I_{DS}$ continues to rise significantly with increased drain bias $V_{DS}$.

To circumvent these SCEs, and extend Moore's law for a few more years, the semiconductor industry has departed from classic planar FET technology. In 2012, Intel launched its 22-nm *Ivy Bridge* CPU. Representing the most radical technology shift in five decades, this chip used a nonplanar transistor known as the finFET [4].

A finFET is built around a thin ridge of silicon, of order 100 nm in total height—like the dorsal fin on a fish's back. Carriers flow from source to drain along this fin. In Sect. 2.3, we derive its first-order *I-V* characteristics. In Sect. 2.4, we look at its layout. In Sect. 2.5, we explore its short-channel behavior.

Before delving into the details, though, let us briefly look at a nonplanar device that is more symmetric and thus easier to understand at a conceptual level. In Sect. 2.2, we examine the basic operation of the *nanowire* field-effect transistor.

## 2.2 The Ideal Transistor

For purposes of maximizing control of the gate over the channel, and avoiding SCEs, the ultimate geometry for an MOS transistor is perhaps a silicon nanowire [5]. Figure 2.2 shows a conceptual view of an *n*-type nanowire FET. Its thickness (or diameter) is $t_{nw}$. Its length, from source to drain, is $L$. Its effective width, $W_{eff}$, is the wire's circumference. Silicon nanowires can be fabricated by conventional MOS masking and etching on bulk wafers [6]. For simplicity, this figure omits such process-specific details as supporting silicon pillars at each end of the nanowire.

Because the gate electrode wraps *all around* the underlying silicon nanowire, the electrostatic control of $V_{GS}$ over the drain-to-source current flow $I_{DS}$ is maximal.

Encroachment by the source-drain depletion regions is minimal, and SCEs have less impact. Source-drain junction depth ($r_j$) plays less of a role. As a result, further scaling down of $L$ can proceed more easily than in the traditional *n*MOS transistor.

### 2.2.1 *Qualitative Behavior of a Nanowire*

Let us investigate the qualitative behavior of this silicon nanowire (SNW) FET. By briefly examining this highly symmetric SNWFET, we will be better able to analyze the less-symmetric finFET structure, covered in detail in the next section.

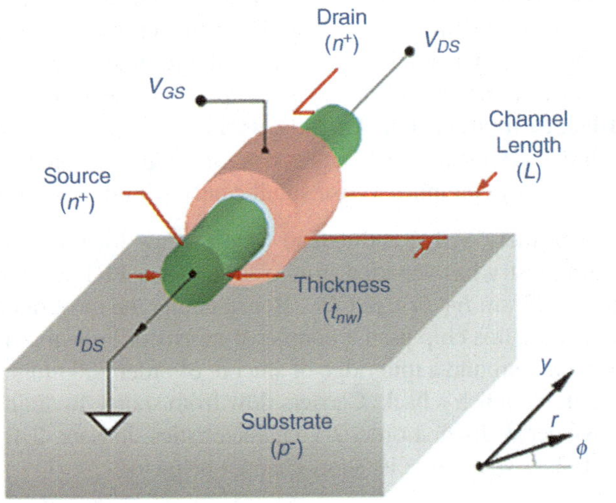


**Fig. 2.2** Conceptual *n*-type nanowire FET (perspective view)

As the applied gate voltage $V_{GS}$ increases, the outer surface of the wire goes through three successive stages: accumulation, depletion, and inversion. We will focus mostly on the latter stage, in which the transistor is fully turned on, and a thin *inversion layer* of mobile electrons carries a current from the source to the drain.

It is natural to use cylindrical coordinates, as in the lower right corner of Fig. 2.2. Coordinate $r$ is the radial distance to any point, perpendicular to the $y$-axis. The coordinate $\phi$ is the angle described as the radius sweeps out a full circle.

Clearly, this transistor is *symmetric* about the $y$-axis. Electrical conditions inside the nanowire—like inversion charge density—are independent of angle $\phi$.

For tutorial purposes, we limit our discussion to thicker nanowires, in which the inversion charge resides in a thin layer at the surface of the wire, at a radius ½ $t_{nw}$. (Thin nanowires cannot be analyzed with the equations of classical physics alone, since electrons confined to nanoscale structures obey the laws of wave mechanics.)

**Physical Insight: Quantum-Mechanical Effects**

A truly small-geometry nanowire exhibits quantum-mechanical effects (QMEs). If the wire diameter $t_{nw}$ is thin (say, below 10 nm), then electrons in the inversion layer are tightly confined in the radial ($r$) and angular ($\phi$) directions. They are only free to drift along the axial direction $y$. Such localization leads to quantization [7].

The tightly confined electrons assert their wave nature and behave like a 1-D electron gas. This causes band splitting and volume inversion. Under volume inversion, the carriers can no longer be viewed as residing in a thin surface layer.

Small-geometry nanowires exhibiting QMEs are discussed in the literature [8]. Analysis starts by solving Poisson's equation, in classical physics, to obtain the electrostatic potential $\Phi(r,y)$ for electrons within the wire. Substituting $\Phi$ into Schrödinger's wave-mechanical equation yields the electron wave function $\psi(r,y)$. The probability density $|\psi|^2$ is used to compute the concentration of electrons [9].

Now let us briefly investigate the three modes of nanowire transistor operation:

1. **Accumulation:** For $V_{GS}$ less than 0 V, the gate has a negative charge. It is balanced by an equal and opposite positive charge on the wire, supplied by majority-carrier holes from the $p$-type interior. These holes accumulate near the surface. The channel region has an excess of holes—but few electrons. With no electrons to flow from $n$-type source to drain, the transistor is an *open circuit.*
2. **Depletion:** In this mode, $V_{GS}$ is positive but still below the threshold voltage for an $n$-type device. Positive charge builds up on the gate, balanced by an equal and opposite negative charge on the nanowire. This negative charge arises as holes in the $p$-type wire are driven away from the positive gate. Each hole is filled with a valence electron, leaving a fixed, negatively-charged acceptor ion.

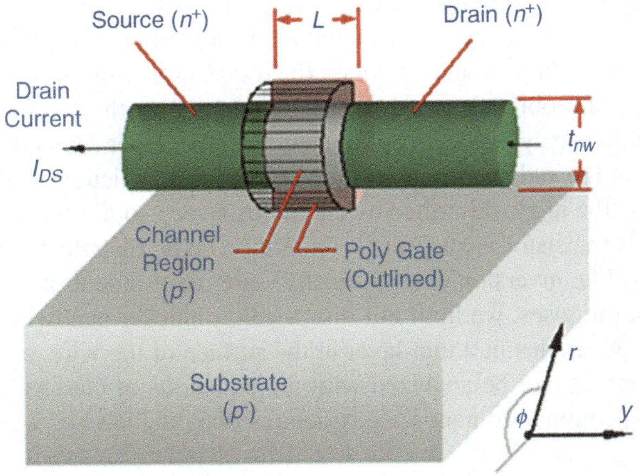


**Fig. 2.3** Conceptual $n$-type nanowire FET (side view)

**Fig. 2.4** Charge profile inside nanowire FET

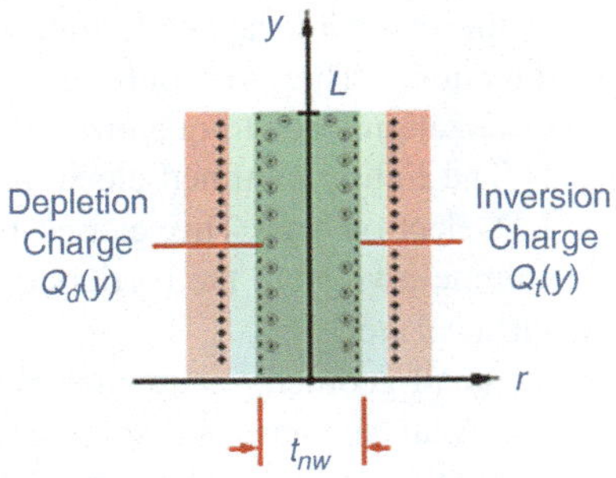


Few electrons exist in the $p$-type wire under the gate. The wire surface is thus depleted of mobile carriers. The transistor is still an open circuit. We will denote this immobile, negative depletion-layer charge (per unit surface area) as $Q_d(y)$.

3. **Inversion:** As $V_{GS}$ exceeds the threshold voltage $V_{Tn}$ for an $n$-type device, inversion sets in. Mobile conduction electrons flood into the nanowire from the source, until these minority carriers *invert* the original doping. The wire surface is now $n$-type instead of $p$-type. This thin inversion layer, a sleeve-like channel extending along the $y$-axis, forms a *conducting bridge* from source to drain. We will denote this mobile inversion channel charge (per unit surface area) as $Q_i(y)$.

Figure 2.3 is a side view of the SNWFET. This cutaway view emphasizes that, during CMOS processing, the channel region inside the wraparound gate electrode remains lightly $p$-type or even undoped (intrinsic). Only the exposed source and drain regions are implanted heavily $n$-type (or $p$-type, for a complementary device). We assume for simplicity there are no trapped charges within the insulating oxide.

Our focus is inversion mode, when the SNWFET is fully turned on. Figure 2.4 is a qualitative plot of charge density along the inverted nanowire. Both components,

$Q_d$ and $Q_i$, are indicated. The profile shown is *uniform* along the entire $y$-axis. This uniform-channel condition exists if the drain bias $V_{DS}$ is small compared to VDD.

Both charge components will play a role in determining the *I-V* characteristics. The inversion charge density $Q_i(y)$ consists of mobile electrons, residing as close to the nanowire surface as they can get. This corresponds to a radial distance ½ $t_{nw}$. The depletion charge density $Q_d(y)$ is made up of $N_A$ ionized acceptor ions per unit volume, at fixed sites throughout the lightly doped nanowire underneath the gate.

This depletion charge forms a layer of finite depth $d$. This depth depends upon process parameters such as $\varepsilon_{Si}$ and $N_A$. (In a thin, lightly doped nanowire, the depth of depletion can reach ½ $t_{nw}$. Such a nanoscale structure is called *fully depleted.*)

In the next section, we briefly consider what happens at higher drain bias. Then the drain end of the nanowire is held at $V_{DS}$, while the source end is at 0 V. We will have to account for the variation of the inversion charge $Q_i(y)$ and the depletion charge $Q_d(y)$ as a function of the distance $y$ from the source terminal.

### 2.2.2 *Potential Distribution in a Nanowire*

Even in as symmetric a structure as the SNWFET shown in Figs. 2.2 or 2.3, it is difficult to find the exact electric field distribution as a function of $r$ and $y$. In this subsection, we explore the complexity of the problem. We then fall back upon the classic simplification known as the *gradual-channel approximation* (GCA).

Instead of deriving the electric field, a vector, it is easier to find the electrostatic potential distribution $\Phi(r, \phi, y)$, a scalar, related to the field by $\boldsymbol{E} = -\nabla\Phi(r, \phi, y)$.

In electrostatics, a scalar potential $\Phi$ arising from a distribution of charge obeys Poisson's equation. Expressed in cylindrical coordinates, it takes the form of (2.1). By cylindrical symmetry, we can omit the derivative term with respect to the angle $\phi$. The volume charge density $qN_A$ on the RHS is applicable to the case of subthreshold operation, $V_{GS} < V_{Tn}$, in which the nanowire is depleted of carriers:

$$\frac{\partial^2\Phi}{\partial r^2} + \frac{1}{r}\frac{\partial\Phi}{\partial r} + \frac{1}{r^2}\frac{\partial^2\Phi}{\partial\phi^2} + \frac{\partial^2\Phi}{\partial y^2} = \frac{qN_A}{\varepsilon_{Si}} \tag{2.1}$$

This still leaves a partial differential equation in two variables, a complex task to solve analytically for $\Phi(r, y)$, unless we resort to simplifying assumptions [10].

To understand these simplifications, let us refer to Fig. 2.5, a conceptual plot of the potential profile $\Phi(r, y)$ inside the SNWFET, at a value of $y$ between 0 and $L$. The vertical axis represents electrostatic potential. Working our way inward, we first see that the entire gate electrode is an equipotential region held at voltage $V_{GS}$.

Inside the thin gate dielectric, $\Phi$ falls steadily as $r$ decreases, corresponding to a relatively high electric field $\boldsymbol{E}_r$. Inside the nanowire, $\Phi$ continues to fall. Its value at the very surface of the silicon—the *surface potential*—is denoted $\Phi_s$. As part of the gradual-channel approximation, we assume that this silicon surface potential $\Phi_s$ remains relatively constant with rising gate bias $V_{GS}$ once inversion is reached.

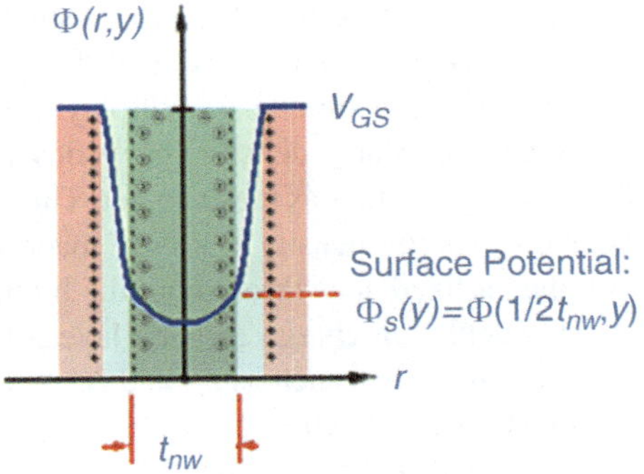


**Fig. 2.5** Potential profile inside nanowire FET

The surface potential is, however, directly affected by increasing drain bias $V_{DS}$. An applied drain bias raises the surface potential at a point $y$ along the nanowire—especially toward the drain end. We thus replace $\Phi_s$ by the sum $\Phi_s + V(y)$, where $V(y)$ reaches $V_{DS}$ at the drain but is 0 V at the source. To first order, the increase is *additive*. This is valid under our assumption of a *gradual* channel, in which the field $\boldsymbol{E}_y$ and potential $\Phi(r, y)$ vary only gradually along $y$ and where $|\boldsymbol{E}_r| >> |\boldsymbol{E}_y|$.

We will rely on this gradual-channel simplification in the subsections ahead, as we investigate the tri-gate finFET. In this subsection, our goal was to introduce the highly symmetric SNWFET, to better understand the less-symmetric finFET later.

The nanowire transistor is itself, however, the subject of intensive research and development. One issue still to be resolved is whether nanoscale wires, as they are scaled down further, can support adequate current flow for digital applications. Another issue is whether the surface roughness encountered by electrons in the inversion layer will be detrimental to their mobility. If these issues are resolved, SNWFETs could well become the future digital logic technology of choice.

Nanowire devices may also hold promise as on-chip *sensors*. Their nanoscale volume, $\frac{1}{4}\pi(t_{nw})^2L$, renders them highly sensitive to biochemical conditions, like minuscule variations in *pH*. Their transistor capabilities may enable such devices to combine an active sensing role with electrical amplification and conversion [11].

**Physical Insight: Poisson's Equation**

Poisson's equation is based upon Gauss' law in differential form, which states: $\varepsilon\nabla\cdot\boldsymbol{E} = \rho$, for a volume charge density $\rho$ in a medium of uniform permittivity $\varepsilon$. If the electric field is replaced with $\boldsymbol{E} = -\nabla\Phi$, where $\Phi$ is the electrostatic potential, then the result is a partial differential equation in three dimensions: $\nabla^2\Phi = -\rho/\varepsilon$.

When applied to the charge distribution in a silicon substrate, nanowire, or fin, the electrical permittivity of crystalline silicon, $\varepsilon_{Si}$, must be used for $\varepsilon$. It is roughly $12\varepsilon_0$, or $12\times$ the permittivity of free space. The factor 12 is the *relative* dielectric constant. A key process parameter, the relative dielectric

(continued)

constant of the channel material, is represented in the BSIM-CMG model by the SPICE parameter EPSRSUB.

Though silicon is a semiconductor, and not a dielectric insulator, its permittivity enters into Poisson's equation because the silicon lattice is highly *polarizable*. Any electric field arising from the charge distribution $\rho$ is inevitably reduced because it polarizes the silicon atoms. The resulting atomic dipoles partially counteract $\boldsymbol{E}$.

It is of historical interest to note that the differential equation developed by the French mathematician Poisson in 1813 is still in use two centuries later as the basis for modeling the most advanced MOS field-effect transistors, including nanowires.

In this section, we briefly investigated an almost ideal transistor, the SNWFET. Building on this conceptual groundwork, let us now examine a nonplanar transistor that is less ideal but is representative of devices already in production at foundries around the world. In the next section, we discuss the bulk tri-gate finFET device.

## 2.3 FinFET *I-V* Characteristics

All the members of the finFET family of devices—whether fabricated on a bulk or SOI substrate, or structured as a double- or triple-gate or a nanowire transistor—share one fundamental characteristic: their operating mechanism is the *field effect*. Drain current flow is under the control of a gate electrode that does not even touch the silicon substrate but—through the influence of a transverse electric field across the gate oxide—*modulates* the conductivity of the underlying fin or nanowire.

Figure 2.6 shows an *n*-type finFET on a bulk silicon substrate. Though generic, it is representative of devices already in production—such as Intel's *Ivy Bridge* die tri-gate transistor [4]. The gate stack can be heavily doped or silicided polysilicon with an $SiO_2$ dielectric, or else an HKMG stack with a refractory metal gate using W, Ti, Mo, or their alloys [12]. For tutorial purposes, we assume the former, with an oxide capacitance (per unit area) of $C_{ox} = \varepsilon_{ox}/t_{ox}$. The silicon fin itself is integral to the underlying substrate. We will assume that it is lightly doped *p*-type.

For the rectangular finFET geometry of Fig. 2.6, it is natural to use the *x*-*y*-*z* coordinates at the lower right. Though not as symmetric as the nanowire transistor of Sect. 2.2, the finFET nevertheless behaves in a similar manner. The silicon fin of thickness $t_{fin}$ plays a very similar role to that of the nanowire of diameter $t_{nw}$.

Let us derive qualitative *I-V* characteristics for this device, from first principles.

We seek to demonstrate that the current $I_{DS}$ flowing through the fin is indeed controlled by the transverse electric field and hence by the applied gate voltage $V_{GS}$. It is also dependent on the applied drain voltage $V_{DS}$. We will again rely on the *gradual-channel approximation* (GCA) introduced in Sect. 2.2. We will bypass algebraic complexity, arriving at an intuitive description of finFET characteristics.

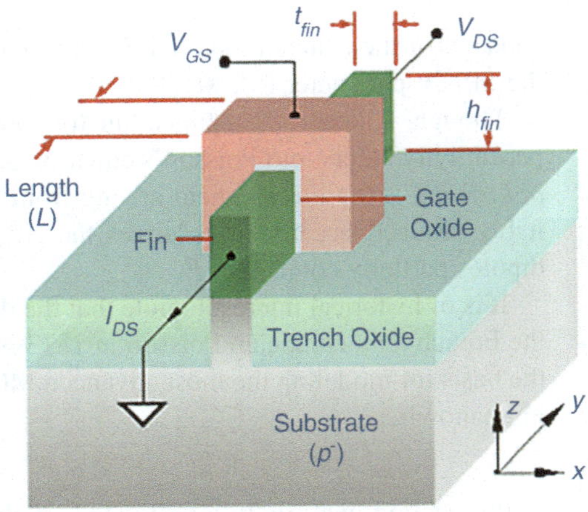


**Fig. 2.6** Generic bulk *n*-type finFET (perspective view)

### 2.3.1 *Uniform-Channel Conditions*

The key observable in MOS transistors is the drain-to-source current $I_{DS}$ as it varies with bias voltages $V_{GS}$ and $V_{DS}$. The *p*-type substrate is assumed grounded. (Upper-case subscripts are used here to denote *large-signal* voltages and currents.)

As in our earlier analysis of the nanowire transistor, there are three finFET bias regions: accumulation, depletion, and inversion. To derive the *I-V* characteristics of a turned-on transistor, we focus on *inversion*. In this mode, the gate voltage $V_{GS}$ exceeds the threshold voltage $V_{Tn}$ for an *n*-type device. A positive charge builds up on the gate. It is balanced by an equal and opposite negative charge on the underlying fin. We will denote this total negative charge (per unit area) as $Q_{fin}$. As in Sect. 2.2, the negative fin charge arises from two independent contributions:

1. A layer of depth $d$ of depletion charge $Q_d$ (per unit area). It consists of acceptor atoms, at uniform concentration $N_A$ throughout the fin, which have been ionized. This charge increases as the depletion layer *deepens*, exposing more acceptors.
2. A thin layer of inversion-channel charge $Q_i$ (per unit area). It consists of mobile electrons flooding into the fin above threshold and residing close to its surfaces.

Summing up both these independent sources, the total negative charge (per unit area) above threshold, built up on all three exposed surfaces of the silicon fin, is:

$$Q_{fin} = Q_i + Q_d \tag{2.2}$$

We assume in this subsection that $V_{DS}$ is *small* compared to the supply voltage VDD. Thus, charge densities do not vary significantly with $y$. Conditions along the length of the channel are *uniform*, from the source at $y = 0$ to the drain at $y = L$.

The silicon fin forms one plate of the capacitor. But *not all* of the applied gate voltage appears across the oxide dielectric. Some gate voltage is dropped between the surface of the fin and its interior. We again denote this *surface potential* as $\Phi_s$:

$$V_{GS} = \frac{-Q_{fin}}{C_{ox}} = \Phi_s \tag{2.3}$$

For a uniform channel, we can assume that surface potential $\Phi_s$ is independent of $y$. Note that we ignore nonideal behavior, such as charges trapped in the oxide layer or at the fin interface. *All* of the fin charge is thus generated by the gate bias.

Rewriting (2.3), and substituting (2.2), we obtain the inversion charge density. For simplicity, we also ignore any process-specific differences in work functions:

$$-Q_i = C_{ox}[V_{GS} - \Phi_s] + Q_d \tag{2.4}$$

We rearrange (2.4) slightly to group together two terms in parentheses that are more dependent upon process parameters such as $N_A$ than on the applied gate bias:

$$-Q_i = C_{ox}[V_{GS} - (\Phi_s - Q_d/C_{ox})] \tag{2.5}$$

As in subsection 2.2.2, we assume that $\Phi_s$ remains relatively constant as $V_{GS}$ is increased, once inversion has been reached [13]. We can identify the parenthesized expression as a fixed *threshold voltage*, dependent on process parameters like $N_A$, $\varepsilon_{ox}$, and $t_{ox}$. When $V_{GS}$ is at the threshold voltage for an $n$-type transistor, the charge $Q_i$ is zero. To a first approximation, the threshold voltage can thus be expressed as:

$$V_{Tn} = \Phi_s - Q_d/C_{ox} \tag{2.6}$$

Using the expression in (2.6), we simplify (2.5) to obtain the inversion-channel charge (per unit area) $Q_i$. It is proportional to the gate voltage *above threshold*. Based on our assumption of a uniform channel, it does not vary appreciably with $y$:

$$-Q_i = C_{ox}[V_{GS} - V_{Tn}] \tag{2.7}$$

As $V_{GS}$ increases further, *above* threshold, more negative charge must appear on the fin. Up to this threshold, the incremental charge was supplied by a deepening of the depletion layer, exposing more immobile ions. But $Q_d$ grows *slowly* with gate bias, only as the square root. Above the threshold, the inversion charge $Q_i$ grows *exponentially* with gate bias, flooding the fin with mobile electrons. It is this abrupt shift in the source of charge that leads to the concept of a *threshold* voltage $V_{Tn}$ [14].

To first order, the depletion layer grows *no deeper* above threshold but remains fixed at a maximum depth $d_{max}$. This will simplify our analysis considerably. All of the second-order effects of depletion charge on *I*-*V* characteristics are lumped into a process-dependent parameter, the threshold voltage $V_{Tn}$. Depletion simply *defers* inversion, until $V_{GS}$ exceeds a value largely determined by the substrate doping $N_A$.

Our simplification does have physical validity. In the limiting case of a fin that is *fully depleted*, $Q_d$ does reach a limit, set by $\frac{1}{2}qN_A t_{fin}$. In exchange for simplicity, we sacrifice the ability to model SCEs. We are not accounting for the deepening of the depletion layer around the drain at higher $V_{DS}$ and its encroachment on the channel. But our model is still adequate to yield qualitative finFET characteristics.

Based on this model, we can derive a useful rule of thumb. As electrons flood into the fin above $V_{Tn}$, they create a thin layer of inversion-channel charge (per unit area) $Q_i$. This conductive layer *bridges the gap* between the source and the drain.

The inversion electrons are *mobile*. Even under a small drain bias $V_{DS}$, they will drift along the channel in ohmic fashion, leaving the source and entering the drain. A microscopic form of Ohm's law states that the average drift velocity of free electrons moving through a silicon lattice is directly proportional to the applied electric field. The proportionality constant $\mu_n$ is the electron mobility. Therefore:

$$\bar{v}_y = -\mu_n E_y = \mu_n |E_y| \tag{2.8}$$

The minus sign arises because electrons move *against* the electric field. On average, the drifting electrons will move from the source to the drain in a time interval $\tau$. This *transit time* is just the length of the channel divided by the average velocity:

$$\tau = \frac{L}{\mu_n |E_y|} = \frac{L^2}{\mu_n V_{DS}} \tag{2.9}$$

where the field $|E_y|$ is of the order of the drain bias $V_{DS}$ over the channel length $L$.

It is this transit time $\tau$ that sets an inherent limit on the speed of the entire technology [15]. An *n*-type transistor in the pull-down network of a CMOS logic gate cannot discharge the load capacitance of the next gate any faster than the time it takes for the required charge to drift across the channel length $L$ of the pull-down device. Though optimistic, this simple delay metric leads us to a first *rule of thumb*.

The transit time $\tau$ depends primarily on device length $L$. As device dimensions are scaled down by a factor of ~$\sqrt{2}$ from one technology node to the next, the density of gate logic on a silicon die scales up by a factor of 2. This is Moore's law. But Eq. (2.9) suggests that the raw speed of the technology will increase by the same factor (partly offset by voltage scaling). This accounts for the industry-wide driving force behind decades of Moore's law: scaling down channel length $L$—even at the cost of higher process complexity—enhances IC operating speed.

**Rule of Thumb 1**

For the nonplanar finFET, just as for planar MOSFETs, the inherent delay of the device *decreases* as the channel length $L$ is scaled down—obeying Moore's law.

In this subsection, under uniform-channel conditions, we can see from (2.8) that $I_{DS}$ grows linearly with the field $\boldsymbol{E}_y$ and in turn with $V_{DS}$. At low drain bias, the ideal finFET thus behaves like a resistor. Here, *low* implies that $V_{DS}$ is sufficient for current to flow yet small enough compared to $V_{GS}$ that the drain bias does not yet affect the inversion charge $Q_i$. In the next subsection, we consider *higher* drain bias.

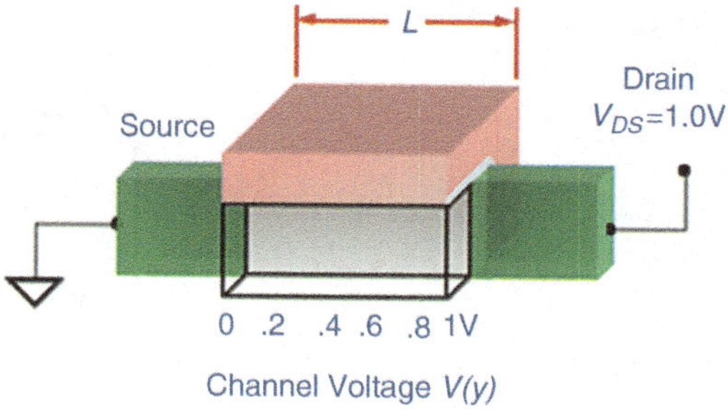


**Fig. 2.7** Channel voltage along the fin (cutaway view)

### 2.3.2 *Gradual-Channel Conditions*

Based on experimental *I-V* plots, we know that $I_{DS}$ does *not* continue to increase linearly with $V_{DS}$. Instead, it will level off—or *saturate*. Because the drain end of the fin is held at $V_{DS}$, less of the applied gate bias will fall across the oxide there. $Q_i$ will thin out near the drain. We must therefore modify Eq. (2.4) to account for the effect of $V_{DS}$ on $Q_i$ along the channel. We thus replace $Q_i$ with $Q_i(y)$. In the interests of simplicity, however, we will assume this charge variation is *gradual*.

Consider Fig. 2.7. Suppose that $V_{DS}$ is 1.0 V. At one end of the fin, the source terminal, the drain bias is 0 V. There, the fin surface potential remains at the value $\Phi_s$ it had at threshold. At the other end, the drain terminal, the applied bias is $V_{DS}$. The surface potential there will be $\Phi_s + V_{DS}$. At any intermediate point $y$ along the fin, the surface potential must take on successive values, like those in the figure.

The ohmic inversion layer, which bridges the gap between the source and drain, behaves in effect like a *resistive* capacitor plate [16]. We can now approximate the potential as a function of $y$. For a gradual channel, in which the gate bias has more influence on the inversion layer than does drain bias, we simply assume the effect is additive [17]. We replace potential $\Phi_s$ at the fin surface with the sum $\Phi_s + V(y)$.

Two boundary conditions must be satisfied at either end of this resistive plate: $V(0) = 0$ V at the source terminal and $V(L) = V_{DS}$ at the drain terminal.

Replacing $\Phi_s$ in Eq. (2.4) with the sum $\Phi_s + V(y)$, we obtain a modified Eq. (2.10) for $Q_i(y)$. Since $V(y)$ must increase with $y$, the inversion charge will thin out toward the drain. This first-order approximation to the effects of drain bias on $Q_i$ is comparable to the classic SPICE Level 1 model for a planar MOSFET:

$$-Q_i(y) = C_{ox}[V_{GS} - \Phi_s - V(y)] + Q_d \tag{2.10}$$

The potential inside the fin is, in reality, a function of the form $\Phi(x,y,z)$. It can only be found accurately by solving Poisson's equation in three dimensions [18]. Advanced SPICE models, like BSIM-CMG Level 72, solve Poisson's equation for a specific set of device parameters using sophisticated analytical approximations [19].

We will proceed, as in Subsection 2.3.1, to bypass mathematical complexities by treating $\Phi_s$ as constant above inversion. We thus lump all the second-order effects of

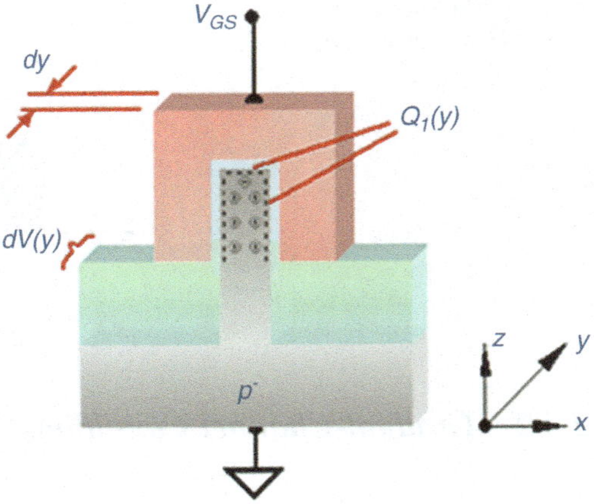


**Fig. 2.8** Elemental channel cross section *dy*

depletion charge buildup into a fixed threshold voltage $V_{Tn}$. The approximate inversion-channel charge (per unit area) as a function of $y$ for a gradual channel is:

$$-Q_i(y) = C_{ox}[V_{GS} - V_{Tn} - V(y)] \tag{2.11}$$

Equation (2.11) is only slightly more complex than (2.7). Let us reemphasize that (2.11) is strictly valid only for long-channel devices, under *gradual*-channel conditions. It cannot model SCEs. But it is adequate for deriving qualitative finFET characteristics and for demonstrating their dependence on IC process parameters.

Let us now compute the drain current $I_{DS}$ flowing through this nonuniform, but gradual, channel. Consider an elemental cross section of the fin, as diagrammed in Fig. 2.8. During a steady-state flow, the current must be the same at any point $y$.

The voltage drop $V(y)$, however, increases along the channel. The differential voltage drop across this section is $dV(y)$. The mobile inversion charge of (2.11) is spread in a thin layer over the three exposed surfaces of the fin (as indicated by tiny minus signs). The fixed depletion charge (circled minus signs) is indicated as well.

Only mobile charge contributes to the drain current. Recalling that $Q_i$ denotes inversion charge *per unit area*, we can express the total mobile inversion charge on all three-fin surfaces, within this elemental volume, as $dq = Q_i\ (2h_{fin} + t_{fin})\ dy$.

Since the carriers in an *n*-type device are negatively charged electrons, drifting along $y$ from source to drain, let us first compute the *source-to-drain* current $I_{SD}$. The source-to-drain current flowing along $y$ in the figure is the charge per unit time drifting through this fin cross section. Since *dy/dt* is the drift velocity, we get:

$$I_{SD} = \frac{dq}{dt} = \frac{Q_i(2h_{fin} + t_{fin})dy}{dt} = Q_i(2h_{fin} + t_{fin})\bar{v}_y \tag{2.12}$$

From Eq. (2.8), the drift velocity is related to the lateral electric field $\boldsymbol{E}_y$ along the fin. But the $y$-component of the lateral field is the negative gradient of the potential $V(y)$. With this substitution in (2.12), we get the differential equation:

$$I_{SD} = \mu_n (2h_{fin} + t_{fin}) Q_i \frac{dV(y)}{dy} \tag{2.13}$$

Equation (2.13) describes drift current only. We neglect the diffusion current of these majority-carrier electrons; this is addressed by more sophisticated models [20]. We can now substitute (2.11) for $Q_i(y)$. Reversing the sign to correspond to a conventional drain-to-source current, we arrive at the simple differential equation:

$$I_{DS} = \mu_n C_{ox} [V_{GS} - V_{Tn} - V(y)] (2h_{fin} + t_{fin}) \frac{dV(y)}{dy} \tag{2.14}$$

We can readily solve differential Eq. (2.14) by integrating over the entire length of the channel, from 0 to $L$. Let us multiply both sides by $dy$ and then integrate:

$$I_{DS} \int_0^L dy = \mu_n (2h_{fin} + t_{fin}) C_{ox} \int_0^{V_{DS}} [V_{GS} - V_{Tn} - V(y)] dV \tag{2.15}$$

We are thus *summing* differential voltage drops along the length $L$ of the channel.

After integrating, we obtain the drain current in terms of bias voltages. This *I-V* characteristic is applicable to an *n*-type gradual-channel finFET above threshold, with an inversion layer that extends from source to drain (and is not pinched off). The resulting Eq. (2.16) is clearly an expression of the electric *field effect*:

$$I_{DS} = \mu_n C_{ox} \frac{(2h_{fin} + t_{fin})}{L} \left[ (V_{GS} - V_{Tn}) V_{DS} - \frac{1}{2} V_{DS}^2 \right] \tag{2.16}$$

The gate voltage $V_{GS}$ creates a transverse field across the oxide, described by $\boldsymbol{E}_x$ and $\boldsymbol{E}_z$. This field modulates the conductivity of the fin. The degree of modulation depends on process parameters like the mobility $\mu_n$ and capacitance $C_{ox} = \varepsilon_{ox}/t_{ox}$.

An equation similar to (2.16) holds for complementary *p*-type finFET devices. In this case, positively charged *holes* flood into the inverted channel, forming a bridge from source to drain. We simply substitute hole mobility $\mu_p$ and the *p*-type device threshold $V_{Tp}$ (a negative value), to obtain the *I-V* characteristic for *p*-type devices.

Figure 2.9 is a conceptual TCAD simulation of an *n*-type finFET, as it carries current. It shows the electron density, profiled at mid-channel, in a lightly doped

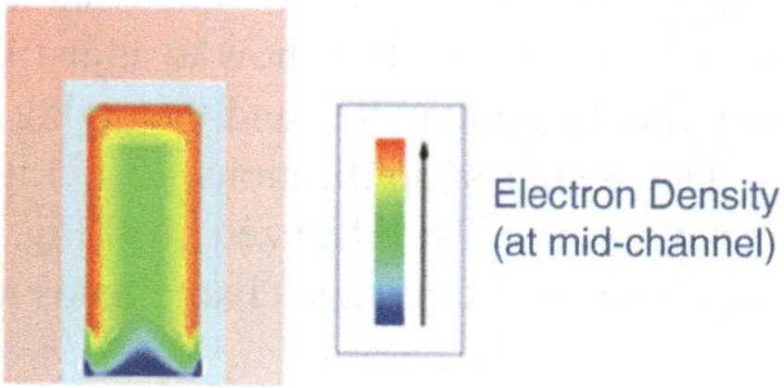


**Fig. 2.9** Conceptual TCAD profile of fin channel

15-nm-thick rectangular fin. Such profiles are obtained using TCAD tools such as Synopsys *Sentaurus Device* [21]. Notice the high density at the three-fin surfaces.

In deriving the *I-V* characteristic of (2.16), we relied upon the gradual-channel approximation (GCA). This is accurate for *long-channel* devices. It loses validity for a *short-channel* device, in which the drain bias $V_{DS}$ applied across a short channel $L$ leads to high lateral field. Such a channel would no longer be gradual, and it would exhibit the same short-channel effects that were cited in Sect. 2.1.

It is important to realize, however, that—within its range of validity—the GCA model faithfully describes qualitative finFET behavior. Even sophisticated SPICE models for finFET devices—like the common-multigate BSIM-CMG algorithm—are founded upon a simplified *core model.* This core model typically starts from GCA assumptions [22]. This ensures the *scalability* of the model. The same algorithm can handle long-channel or short-channel finFETs, without encountering discontinuous currents or voltages—and without the need for artificial tuning parameters [23].

The current flow in (2.16) depends on channel length $L$, a layout parameter, just as it does for planar MOSFETs. Notice, however, a radical departure from planar technology: no layout parameter for channel width $W$ appears in the equation. Instead, the finFET device has an *effective* width fixed by two process parameters:

$$W_{eff} = \left(2h_{fin} + t_{fin}\right) \tag{2.17}$$

We can visualize the thin inversion layer as *folded over* the exposed fin surfaces. Its extent lies in *two* dimensions, $x$ and $z$. The finFET is indeed a nonplanar device.

**Rule of Thumb 2**

The nonplanar finFET exhibits the same qualitative $I_{DS}$ dependence on $V_{GS}$ and $V_{DS}$ as did the traditional planar MOSFET, under gradual-channel conditions, provided that the planar width $W$ is replaced by the nonplanar effective width $W_{eff}$ in (2.17).

This qualitative similarity has been borne out both by SPICE simulations and by experimental results [24]. Though its *I-V* characteristics, according to (2.16), follow the same first-order behavior as did the planar MOSFET, the nonplanar finFET is inherently more immune to SCEs. In Sect. 2.5, we will explore the reasons why.

Although we did not derive an *I-V* characteristic for the nanowire transistor of Sect. 2.2, it can be shown that it obeys, under GCA assumptions, an equation like (2.16)—except that its effective width is $\pi t_{nw}$, the nanowire's *circumference* [25].

Equation (2.17) revealed that the width of a single-fin transistor is completely fixed by the fabrication process. But manual sizing of transistors is often critical to analog, and even standard-cell, designers, to meet difficult speed/power trade-offs. In Sect. 2.4, we will see how to adjust $W_{eff}$ for an individual transistor by laying out multiple fins in parallel, and then tying them together electrically, in shunt.

Due to our simplifications, (2.17) ignores the potential *corner effects* [26] in a rectangular fin. The idealized fin of Fig. 2.6 has two sharp edges running along the $y$-axis. Just as electrostatic charge tends to accumulate at the *tip* of a teardrop-shaped conductor, any sharp fin edges and corners can lead to excessive channel charge density, accompanied by high electric fields, causing premature inversion.

We have ignored such edge effects by viewing the channel layer as *folded over* the fin. If the corner radius is not carefully controlled during fabrication, a tri-gate finFET can even exhibit *two* threshold voltages. The following insight outlines the common IC processing techniques used to mitigate the effects of sharp fin edges.

**Physical Insight: Double-Gate Hard-Masked FinFETs**
Sharp edges and corners can lead to parasitic inversion channels along the fin. One technique to avoid this problem is to fabricate a thick dielectric layer—termed a *hard mask*—atop the fin [27]. Now the uppermost surface of the fin will never invert and cannot participate in current flow. The hard-masked device effectively has *two* gates, both connected electrically, with one controlling each side of the fin.

The BSIM-CMG model includes a SPICE parameter named `TMASK`. It is used to specify the *thickness* of the hard mask. Setting `TMASK` = 0 indicates that there is *no* hard mask; the dielectric thickness atop the fin then simply defaults to $t_{ox}$.[28].

Another aspect of some finFET structures is a fin having a *rounded top*. This is evident in TEM images of the Intel *Ivy Bridge* tri-gate transistor. It is likely there to reduce intense fields at the corners—and thus to enhance long-term reliability [4].

In such cases, $W_{eff}$ is often approximated by $2h_{fin}$. This applies when $t_{fin} << h_{fin}$. It also applies to double-gate devices, with a hard mask on top of the fin.

### 2.3.3 Pinched-Off Channel Conditions

From Eq. (2.16), it is clear that the current $I_{DS}$ begins to level off at higher $V_{DS}$ values, for fixed $V_{GS}$. This gradual leveling off of the *I-V* characteristic with higher drain bias is known as *saturation*. As we discussed in Sect. 2.3.2, this happens because the inversion charge layer *thins out* toward the drain terminal, where the potential drop across the MOS capacitor is lower, by the applied bias $V_{DS}$.

At a critical drain bias $V_{Dsat}$, the inversion charge $Q_i(y)$ thins out until reaching 0 at the drain end of the fin. The conductive channel is then said to be *pinched off*.

Let us derive the drain current and drain bias at this critical point of saturation. We evaluate Eq. (2.11) at $y = L$. Setting $Q_i(L) = 0$ at pinch-off, and applying the boundary condition $V(L) = V_{Dsat}$, we obtain the pinch-off condition below:

$$V_{Dsat} = V_{GS} - V_{Tn} \tag{2.18}$$

At this drain bias, the channel is pinched off at $y = L$. The potential drop across the MOS capacitor at that point becomes inadequate to support inversion. But remarkably, drain current will continue to flow, even as the drain bias rises higher.

We can compute this saturation drain current by substituting (2.18) into (2.16):

$$I_{Dsat} = \mu_n C_{ox} \frac{W_{eff}}{2L} (V_{Dsat})^2 \tag{2.19}$$

At pinch-off, we reach the limits of our gradual-channel approximation. The slope of $I_{DS}$ in (2.16) becomes zero. We conclude from (2.19) that the drain current has reached a saturated value, $I_{Dsat}$, and no longer increases with drain bias. Though our GCA model loses validity above $V_{Dsat}$, we can explain this current intuitively.

Beyond pinch-off, at drain bias $V_{DS} > V_{Dsat}$, the inversion layer terminates even before $y = L$. It no longer provides a conductive bridge from the source to the drain.

The voltage at the pinch-off point is still $V_{Dsat}$. Beyond it lies a narrow gap, a depleted region, of length $\Delta L$. The voltage drop across this gap is then $V_{DS} - V_{Dsat}$.

The high electric field $|\boldsymbol{E}_y|$ in the narrow gap sweeps electrons into the drain. For a long-channel device, $\Delta L << L$. Under these conditions, the inversion channel is not much shorter and still has the same endpoint voltage $V_{Dsat}$. To first order, the current flowing through it will remain constant at the saturated value in (2.19).

This physical situation has been likened to a waterfall [29]. As the river runs over the rapids, and the water plummets over the falls, its flow rate is determined not by the height of the falls but by the volume of water flowing through the rapids.

More sophisticated models take into account the shortening of the channel above pinch-off. This leads to the *channel-length modulation* (CLM) cited in Sect. 2.1. Especially in short-channel devices, the current slowly *rises* with higher drain bias.

The saturated transistor thus behaves as a constant-current source in shunt with an output resistor. SPICE simulations indicate that the tri-gate finFET has higher output resistance, and thus lower CLM, due to better electrostatic gate control [30].

From Eq. (2.19), we can derive a rule of thumb regarding the *drive current* of a bulk finFET. Drive is usually expressed as the drain current $I_{ON}$ flowing when both $V_{GS}$ and $V_{DS}$ are biased at the supply voltage VDD. With $V_{GS}$ tied to $V_{DS}$, any transistor with current flowing is always *saturated*. Thus, (2.19) applies. Ignoring CLM effects, (2.19) tells us that $I_{ON}$ is determined solely by process parameters.

**Rule of Thumb 3**

For a nonplanar *n*-type finFET, similar to a planar MOSFET, the drive current $I_{ON}$ is proportional to $\mu_n\, C_{ox} W_{eff} / L$. Thus, current drive can be enhanced by scaling down $L$, increasing $W_{eff}$, thinning down oxide thickness $t_{ox}$, or enhancing the mobility $\mu_n$.

For tutorial purposes, we have regarded the mobility $\mu_n$ as a constant, determined by the silicon lattice. This corresponds to the SPICE parameter U0. In sophisticated models like BSIM-CMG, this base parameter is adjusted to account for variations in channel length, temperature, and field—even fin characteristics like surface roughness. Mobility is *degraded* in thin-fin transistors, due to the scattering of charge carriers off the uneven fin sidewalls. Advanced fabrication techniques such as hydrogen annealing *smooth* the fin sidewalls; this is a practical example of Rule of Thumb 3.

Equation (2.19) also reveals that drive current is affected by the oxide dielectric, primarily through the gate capacitance (per unit area) $C_{ox}$. This is a *ratio*, $\varepsilon_{ox}/t_{ox}$.

We can enhance current drive by increasing the numerator, without the need to scale down the denominator. Oxides are already at the limits of scaling—at 4 or 5 molecules thick. Intel introduced the high-κ metal-gate (HKMG) stack, using $HfO_2$ (or hafnium-based) dielectrics, whose permittivity is ~4× higher than that of $SiO_2$.

**Rule of Thumb 4**

The electrical and physical characteristics of finFET devices are enhanced by many of the same process refinements used for planar MOSFETs, including the HKMG gate stack, and mobility improvement by straining the silicon lattice [42, 43].

Now that we have derived the qualitative electrical characteristics of a generic finFET from first principles, let us turn our attention to its *geometric* characteristics.

## 2.4 FinFET Layout

Our goal in this section is to consider the layout of a typical CMOS standard cell comprising both *n*- and *p*-type finFETs. Many traditional features of planar CMOS geometry carry over to nonplanar finFET technology. In particular, the trend toward highly regular *litho-friendly* layout, with uniformly pitched polysilicon gate lines, has become even more essential in nanoscale finFET fabrication processes.

The biggest departure from traditional CMOS layout is that transistor width $W$ is no longer a *design* parameter. Equation (2.17) stated that the effective width $W_{eff}$ of a one-fin device is fixed by *process* parameters: fin height $h_{fin}$ and thickness $t_{fin}$.

### *2.4.1 Fins in Parallel*

To adjust the effective width of a finFET, we must resort to laying out *multiple* fins, in parallel. Figure 2.10 shows a single *n*-type transistor with three fins. They share a common gate, running across all three fins. These three fins, in shunt, carry a total drain-to-source current $I_{DS}$. The effective finFET width has thus been tripled.

Figure 2.10 indicates schematically that the three drain ends are tied together to form a common drain. Similarly, the three source ends are tied together electrically to form a common source. Each fin will thus carry 1/3 of the total current flow $I_{DS}$.

Based on this geometry, we can extend Eq. (2.17) for the effective width of a single finFET by including an integer multiplier $N_{fin}$ (equal to 3 in this figure):

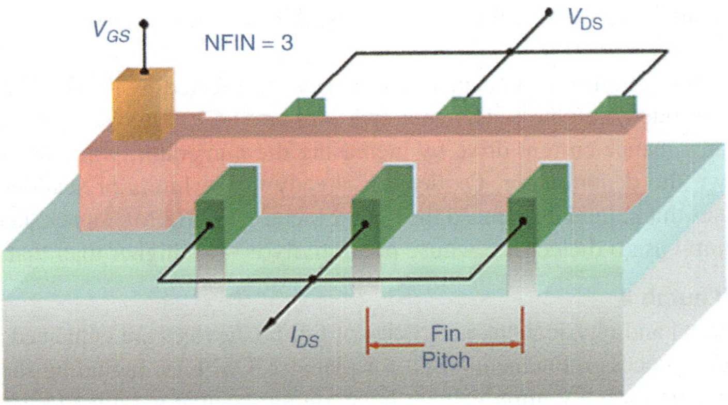


**Fig. 2.10** An $n$-type finFET with three fins (perspective view)

$$W_{eff} = N_{fin}(2h_{fin} + t_{fin}) \tag{2.20}$$

Integer multiplier $N_{fin}$ corresponds to the BSIM-CMG SPICE parameter `NFIN`. In practice it has ranged from 2 to 10. This leads us to a fundamental rule of thumb:

**Rule of Thumb 5**
The effective width $W_{eff}$ of a finFET device is *quantized*. It takes on discrete values, integer multiples of the single-fin width $2h_{fin} + t_{fin}$. The multiplier is denoted $N_{fin}$.

The connections shown schematically in Fig. 2.10 are, in reality, not easy to fabricate. In nanoscale technology, source-drain contact cuts have become so tiny that *contact resistance* threatens to dominate other parasitic parameters. Each fin in Fig. 2.10 has to carry 1/3 of the total current flow. Significant fin-to-fin variations in contact resistance would render the drive current *unpredictable*, thus defeating the very purpose of tying fins together in shunt. In the next subsection, we will look at one process-specific solution—a radical departure from classic silicided contacts.

According to (2.20), a straightforward method of enhancing the drive current is to increase the number of fins. Figure 2.11 shows an $n$-type finFET with *five* fins. Notice that it still occupies the *same silicon area* as the transistor of Fig. 2.10. More fins were packed into the same available width. We thus achieved 5/3 the drive current, for the same silicon area. But this increased drive comes at a price.

Layout designers express the dimensions of on-chip geometry using the metric of *fin pitch*—the center-to-center spacing between adjacent fins. In Fig. 2.10, *two* fin pitches were required to lay out three adjacent transistors. In Fig. 2.11, *four* fin pitches were consumed for five transistors—but only by *halving* the fin pitch.

Half the pitch requires *twice the resolution* during photolithography. This need has been met by—and in turn fuels the demand for—advanced masking techniques such as self-aligned double patterning (SADP) for critical layers like fins. Double patterning uses *two* offset masks to pattern a single CMOS layer, instead of one [43].

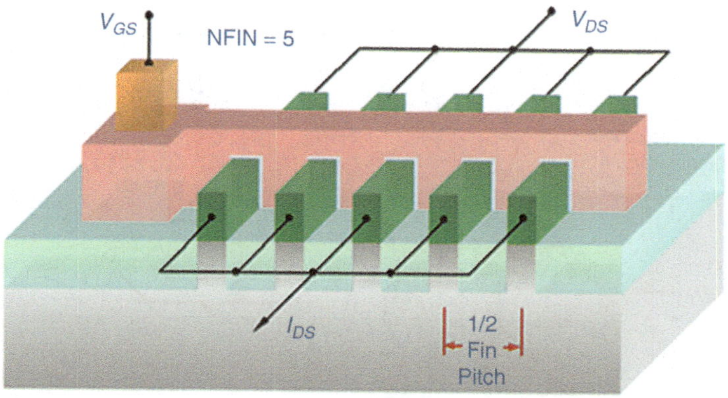


**Fig. 2.11** An *n*-type finFET with five fins (perspective view)

**Rule of Thumb 6**
The need for higher finFET drive, achieved by multiple parallel fins, contributes to an industry demand for self-aligned multiple-patterning photolithography methods.

The next subsection explains how the schematic connections in Fig. 2.10 or Fig. 2.11 are fabricated. We will introduce an *extra metal layer* called METAL0.

**Physical Insight: Litho-Friendly Layout**
The bright-colored patterns visible on the underside of an ordinary compact disc are due to light diffracted from the closely spaced 0.5-μm pits that encode the data.

A mask used in photolithography, to print a layer of geometric features onto the wafer, consists of rectangles and polygons spaced even more closely. A feature size of 22 nm is less than the wavelength of the 193-nm UV light transmitted through the mask. This causes the light to *diffract* (or spread), thus limiting CD resolution.

Resolution enhancement techniques like *optical proximity correction* (OPC) are necessary to compensate for the distorting effects of neighboring shapes. OPC tools are used, post-layout, to modify the GDSII data describing the geometry on a layer. Just as a small typeface with serifs looks clearer, the altered shapes will print better.

But OPC is computationally intensive, and a single layer can easily run overnight.

Litho-friendly layout is a highly regular design style that alleviates the task of OPC. It avoids any jogs or bends, preferring straight lines and a uniform pitch. Nanoscale finFET technology continues this trend. Fins are laid out in parallel, perpendicular to their poly gate. The gates are also laid out in straight parallel lines. To maximize regularity, dummy poly lines, over STI areas, are drawn at cell edges.

Litho-friendly layout enhances chip manufacturability, at the cost of cell area.

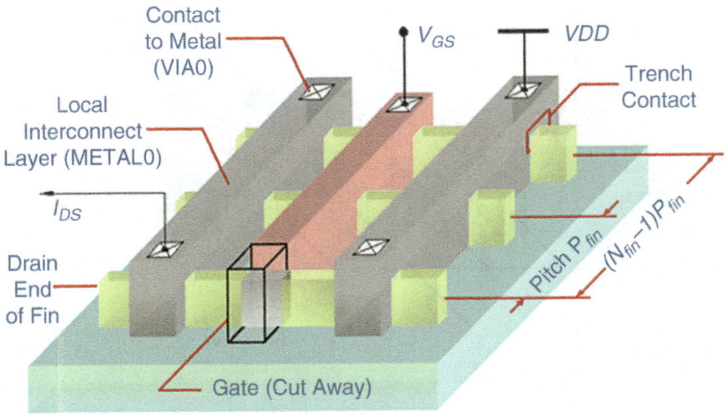


**Fig. 2.12** A *p*-type finFET with three fins (side view)

### 2.4.2 Local Interconnect Layer

Previous figures showed schematically how multiple fins can be tied together to increase the finFET device's effective width. In the wafer fab, making predictable, low-resistance contacts between individual fins and a metal line is a daunting task.

To solve this issue, fabs like Intel, Samsung, and TSMC have departed radically from classic source-drain contact engineering. At 22 nm, a *local interconnect layer* was introduced. It uses lines of a refractory metal, like tungsten, running parallel to the gate. Tungsten (W) is ideal for these local connections, because it is able to fill in tiny spaces more easily than copper. It appears as dark gray bars in Fig. 2.12.

Since this local interconnect layer is below METAL1, it is informally known as METAL0 [43]. It is *localized*, since it does not extend beyond the ACTIVE area of a single transistor. At a silicon foundry, this layer is fabricated after front-end-of-line (FEOL) base layers but prior to back-end-of-line (BEOL) metallization. It is thus informally referred to as a *middle*-of-line (MOL) fabrication step.

Let us illustrate the resulting layout using a *p*-type transistor as our example. Exactly the same layout techniques would apply to a complementary *n*-type device.

Figure 2.12 is a *p*-type transistor with three fins. These fins are all implanted PPLUS (except under the poly gate, as shown by the cutaway view). A METAL0 line at left ties together the *drain* ends. A METAL0 line at the right ties together the *source* ends. They implement the schematic connections in Figs. 2.10 and 2.11.

The common source must connect to the VDD rail, on the METAL1 layer (not shown). To avoid clutter, an × symbol indicates the site of the square contact cut between METAL1 and METAL0. This contact-cut layer is informally called VIA0.

There is a *direct connection* everywhere a METAL0 line crosses a fin. One such connection is outlined by the small rectangle. At that point, a *trench* is etched into the fin. Then an epitaxial silicon apron (not shown) is grown, widening the fin a bit.

The METAL0 line thus solidly contacts the underlying fin on *several* surfaces. This ensures a predictable, low-resistance connection to each individual fin [4].

From Fig. 2.12, we can infer a key advantage of nonplanar finFET geometry. In this side view, the available width is roughly two fin pitches, or $(N_{fin} - 1)P_{fin}$. This represents the width $W$ that a planar MOSFET would have, if laid out over the same footprint. Using (2.20), we get the ratio below, for a multiple-fin transistor:

$$\frac{W_{eff}}{W_{MOS}} \approx \frac{N_{fin}\left(2h_{fin} + t_{fin}\right)}{\left(N_{fin} - 1\right)P_{fin}} \tag{2.21}$$

Let us substitute realistic values into (2.21), based on Intel's second-generation finFET technology, at the 14-nm node. It utilized a fin pitch and a fin height both equal to 42 nm. The fin width was 8 nm [31, 43]. The ratio that results from (2.21), for a three-fin transistor, is over 3. This leads to the rough rule of thumb below:

**Rule of Thumb 7**
Nonplanar finFET devices can pack ~3× the transistor width into the same area as planar devices—since their effective width $W_{eff}$ extends partly in the $z$-direction. This implies high drive, without any impact on circuit area—a critical advantage in such applications as high-density SRAM memory arrays designed using finFETs.

According to (2.21), this advantage hinges on a tight fin pitch $P_{fin}$ and a tall fin of height $h_{fin}$. Fabricating very tall fins, of course, poses difficulties in processing. Fins taller than $4t_{fin}$ may risk instability unless more area is allowed at the base [32].

We have investigated the geometry of a single finFET with multiple fins. Now let us extend these principles to the detailed layout of a typical CMOS standard cell.

### 2.4.3 Standard-Cell Layout

Figure 2.13 shows a generic standard-cell layout of a finFET NAND2 gate. This layout is litho-friendly, arranged along regularly spaced horizontal and vertical lines. The two vertical poly lines are driven by the gate's inputs A and B. Running parallel to these lines are the local interconnects, on METAL0, which tie together the source or drain ends of the fins. (To improve clarity, no dummy poly lines are shown. Neither is the required $n$-type well that must enclose the $p$-type transistors.)

Recall that each METAL0 segment directly connects to the fins that it overlaps. Each such segment thus defines a common source or drain. In turn, the METAL0 segment connects to the next metal layer, METAL1, through at least one VIA0 cut.

An active area defines a rectangular region for transistors of the same type. Outside an active rectangle is STI oxide. The lower active area in Fig. 2.13 has two $n$-type transistors in series, the pull-down network of this NAND2. Output Y is thus pulled down to ground if *both* gate inputs A and B are at a logic-high value.

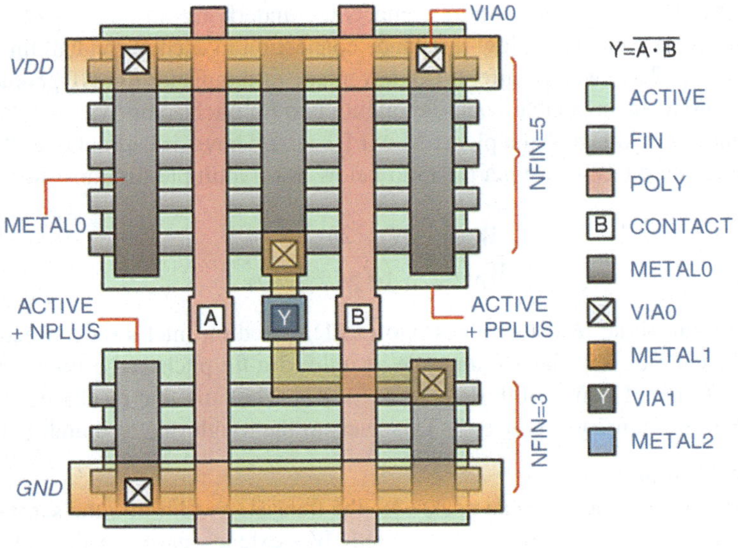


**Fig. 2.13** NAND2 standard cell (layout view)

The upper active area has two *p*-type transistors in shunt—the pull-up network. Output Y is pulled up to the VDD rail if *either or both* inputs are at a logic-low value. The resulting gate output is thus Y = ~(A·B).

Notice that the *p*-type transistors in the pull-up network are laid out with five fins, while the *n*-type devices in the pull-down network have three fins. Transistors are sized in this way to balance rising and falling NAND2 output delays. Increasing the width of *p*-type devices compensates for the lower mobility of holes—equalizing speed.

Mobility engineering in finFETs, however, is a topic of intensive research. Such process enhancements as fin crystal orientation promise simultaneous optimization of $\mu_n$ and $\mu_p$ without the need for the designer to widen every *p*-type transistor [33].

We can conclude from Fig. 2.13 that laying out a standard cell using finFETs is a more complex design task than for planar MOSFETs. Laying out an analog macro can take 4× the effort, although the performance is well worth the effort [34].

Automated layout tools can facilitate the task. With Synopsys *Custom Designer*, for example, the user can draw a finFET transistor to the same width *W* as its planar counterpart—and the tool will split it into multiple fins, at design-rule spacing [35].

Referring to this standard cell, we can infer several geometric rules of thumb:

**Rule of Thumb 8**
An exposed fin, crossing an ACTIVE area, and surrounded by NPLUS (or PPLUS), becomes the source or drain of a transistor. If overlapped by POLY, it is a channel.

**Rule of Thumb 9**
A METAL0 line crossing fins within an ACTIVE area directly ties them together, at the local interconnect level, forming the common source or drain of a transistor.

**Rule of Thumb 10**
The output of a typical (static) CMOS standard cell is a metal segment that connects a source or drain that is inside the NPLUS mask, with one inside the PPLUS mask.

In the previous sections, we introduced nonplanar transistors like the SNWFET and the finFET. We derived their electrical characteristics to first order, using GCA assumptions. Along the way, we listed practical rules of thumb—some of which reflected qualitative similarities between planar and nonplanar field-effect devices.

The next section explores the key difference: a nonplanar transistor of length $L$ is inherently *more immune* to SCEs than a traditional planar device of equal length.

## 2.5 Short-Channel FinFETs

We saw in Fig. 2.1 that a short-channel MOSFET is no longer very planar. The gate fully controls only a *fraction* of the channel length $L$. It loses control over the areas near the source and drain—indicated in that figure by two shaded triangles.

This encroachment arises from the depletion region around the source and drain. At its maximum depth, the total depletion charge is $-qN_AWLd_{max}$. But the shaded fraction of depletion charge in Fig. 2.1 needs *no gate voltage* to build up. It is a by-product of the *p-n* junctions between the source or drain and the substrate. The channel inverts at a *lower* gate voltage, and $V_{Tn}$ *rolls off* with shorter length $L$. With further scaling, the relative fraction of shaded charge grows. Roll-off gets worse.

This roll-off can also be inferred from Eq. (2.6). This expression for $V_{Tn}$, derived under gradual-channel conditions, includes a total depletion charge $|Q_d|$ induced by the gate bias. For shorter channels, a fraction of this charge is built into the source-drain *p-n* junctions and needs no gate voltage. Thus, $V_{Tn}$ becomes lower.

In striking contrast to the traditional planar short-channel MOSFET, nonplanar devices like the SNWFET in Figs. 2.2 and 2.3, and the finFET in Fig. 2.6, are far more symmetric. Their gate electrodes *wrap around* the conducting channel—all the way, for a nanowire device, or three-fourths of the way, for a tri-gate finFET.

This wraparound effect maximizes the electrostatic control of gate over channel. This explains *qualitatively* why the finFET device would show higher immunity to the SCEs cited in Sect. 2.1. But is there a more *quantitative* explanation? Can we find a rule of thumb to estimate the channel length $L$ below which SCEs will set in?

### 2.5.1 *Natural Screening Length*

The answer to these questions lies in the detailed solution of Poisson's equation. We visited this equation in (2.1), for a cylindrically symmetric SNWFET. A similar equation, using rectangular coordinates, applies to the less-symmetric finFET. Its analytical solution is even more complex and well beyond the scope of a tutorial.

**Table 2.1** Natural screening length λ for field-effect devices [39, 40]

| | | |
|---|---|---|
| **a.** Planar bulk MOSFET | $\sqrt{(\varepsilon_{Si}/\varepsilon_{ox})t_{ox}d_{\max}}$ | 22 nm |
| **b.** Nonplanar finFET (tri-gate) | $\sqrt{(\varepsilon_{Si}/3\varepsilon_{ox})t_{ox}t_{fin}}$ | 4 nm |
| **c.** Nanowire FET (cylindrical) | $\sqrt{\dfrac{2\varepsilon_{Si}t_{nw}^2 \ln(1+2t_{ox}/t_{nw}) + \varepsilon_{ox}t_{nw}^2}{16\varepsilon_{ox}}}$ | 3 nm |

Sample values computed for $t_{fin} = t_{nw} = 8$ nm, $t_{ox} = 1.5$ nm, $N_A = 10^{17}$ cm$^{-3}$

Poisson's equation can, however, be transformed into the simpler form (2.22), where λ is a process-dependent *length* parameter and $\varphi(y)$ is the transformed potential (with an added fixed-charge term) along the channel [36–38]:

$$\frac{d^2\varphi}{dy^2} = \frac{1}{\lambda^2}\varphi(y) \tag{2.22}$$

This transformed differential equation, along with its boundary conditions at the source and drain, has a solution made up of exponential terms of form $\exp\pm(y/\lambda)$. Parameter λ is interpreted as a natural *screening length*, with values as in Table 2.1.

Physically, λ is a characteristic length representing the degree of *encroachment* of source-drain field lines into the channel (like the shaded triangles in Fig. 2.1). It depends on device geometry and process parameters. If λ is a large fraction of $L$, then the influence of source or drain over the channel will fall away slowly, due to $\exp-(y/\lambda)$ or $\exp-[(L-y)/\lambda]$ terms. Such a device exhibits pronounced SCEs.

But if λ is a *small* fraction of $L$, then the undesired influence of the source or drain over channel conditions falls away more quickly. Beyond a few λ, the field lines from source or drain become negligible. Such a device will be free of SCEs.

Though λ is highly dependent on process-specific parameters, the sample values in Table 2.1 suggest that the tri-gate finFET has a smaller λ than a planar MOSFET. This explains quantitatively its inherently higher immunity to short-channel effects. (The SNWFET, especially at smaller diameter $t_{nw}$, is potentially the most immune.)

**Rule of Thumb 11**
To shorten the screening length λ for the tri-gate finFET, thus suppressing SCEs, reduce the fin thickness $t_{fin}$ or increase the gate capacitance $C_{ox} = \varepsilon_{ox}/t_{ox}$.

For a tri-gate finFET device, Table 2.1 shows that λ depends only on the *product* $t_{fin}\, t_{ox}$. This allows trade-offs. We can scale down *either* fin thickness or gate oxide. Notice another advantage of the finFET. There is no need to increase the channel doping, just to reduce depletion-layer width $d_{max}$, as in traditional planar devices. Lightly doped fins enhance mobility, since carriers are not scattered as frequently.

Based on the exponential form of solutions to (2.22), and recalling that $\exp(-3)$ is 5%, we can state a rough rule of thumb that applies to any device in Table 2.1:

**Rule of Thumb 12**
To avoid SCEs, a field-effect device should have at least a channel length $L > 3\lambda$.

Referring to the sample values of $\lambda$ in Table 2.1, row **b**, we can conclude that a tri-gate finFET device of channel length 14 nm (or $3.5\lambda$) is relatively free of SCEs. But a planar MOSFET of the same length ($0.6\lambda$, from row **a**) has pronounced SCEs.

**Physical Insight: Natural Screening Length**
Mathematically, a parameter like $1/\lambda^2$ in (2.22) is known as an *eigenvalue* of the differential equation. Derivative $d^2\varphi(y)/dy^2$ returns the original function, scaled by a factor $1/\lambda^2$ whose value is fixed by boundary conditions. This invariance under transformation holds only for specific, physically meaningful values. Schrödinger's equation for an electron inside a potential well, for example, results in quantized energy eigenvalues which define the allowed stationary states of the electron.

Screening length $\lambda$ seems to be a physically meaningful parameter for nonplanar transistors. This has been demonstrated by numerical simulations of short-channel effects (such as DIBL) versus device length $L$. When such simulations are run on a wide variety of double-gate, tri-gate, and similar nonplanar devices—and plotted against the normalized variable $L/\lambda$—they all fall on the same common curve [41].

### 2.5.2 BSIM-CMG SPICE Model

During the preceding sections, we considered several members of the family of nonplanar transistors. Though the geometry of a nanowire device appears radically different from that of a bulk tri-gate or double-gate device, they in fact share many common properties. In this subsection, we take a brief look at an industry-standard model, BSIM-CMG, which can simulate any of these family members using SPICE.

The output plots or data from BSIM-CMG simulations accurately model all of the short-channel effects and quantum-mechanical effects exhibited by these devices.

This turnkey *common multiple-gate* model applies to double-gate, triple-gate, or wraparound-gate devices—whose gates all form part of the same electrical node and share a *common* bias $V_{GS}$. (Devices like a double-gate finFET in which separate bias voltages can be applied *independently* to the two gate halves must use another model, BSIM-IMG. In this tutorial, we have considered only common-gate devices.)

These are complex models. To run a SPICE simulation using BSIM-CMG, users must specify—or accept defaults for—over 200 parameters describing the device and process. Foundries routinely provide this data to their customer base, in the form of $n$-type and $p$-type `.MODEL` files that can be included in a SPICE deck. We will highlight just a few global parameters which determine the type of device.

The `BULKMOD` parameter selects the kind of wafer on which the finFET devices are fabricated. It has two values: 0 for SOI substrates and 1 for bulk silicon substrates. The `GEOMOD` parameter selects the geometry model. The values relevant to this tutorial are 0 for double-gate finFET, 1 for tri-gate, and 3 for cylindrical SNWFET.

The bulk $p$-type five-fin transistor used in the NAND2 of Fig. 2.13 would thus be modeled by a SPICE card with many parameters. The first lines might resemble:

```
.MODEL pFET pMOS LEVEL=72
+ BULKMOD=1 GEOMOD=1 NFIN=1 EPSRSUB=12 U0=0.03 ....
```

where the default fin number is overridden when the transistor is instantiated by specifying `NFIN = 5`. Notice NFIN now replaces the traditional MOSFET width W. The BSIM-CMG model is written in Verilog-A. Developed at UC Berkeley, it can be downloaded at no cost, with a technical manual describing all its parameters [28].

### 2.5.3 Strengths and Weaknesses

In this closing subsection, we summarize a few of the strengths and weaknesses of the generic bulk $n$-type finFET in Fig. 2.6 (or its $p$-type counterpart). Based on published data from Intel, this nonplanar 22-nm tri-gate transistor is in high-volume production [42]. It has a higher $I_{ON}$ and lower $I_{OFF}$ current than did its planar 32-nm predecessor. The higher $I_{ON}/I_{OFF}$ ratio demonstrates improved immunity to SCEs.

In addition, finFET technology has proven versatile, with options for low-power, high-speed, I/O-voltage, and on-chip analog/RF devices. Its first-generation success has led Intel to develop a second-generation finFET with a more rectangular fin [43].

Several of the representative strengths of finFET devices are summarized below:

- **Enhanced scalability:** Due to improved electrostatic control of the gate over the channel—as indicated by smaller values of screening length $\lambda$—the finFET is less prone to SCEs such as threshold roll-off and subthreshold conduction. This enhanced scalability potentially extends Moore's law for years to come.
- **Lower $I_{OFF}$ current:** Below threshold, the finFET is turned off more fully than a planar MOSFET of similar dimensions. This reduces standby leakage power. The figure of merit typically employed is low *subthreshold swing*, measured in mV/decade of leakage current. Intel's 22-nm low-power finFET devices showed a swing of only 65 mV/decade, compared to 100 for planar 32-nm devices [42].
- **Higher $I_{ON}$ drive in same-cell area:** Standard cells using finFET transistors can pack several times the device width into the same cell footprint as planar standard cells—because the effective width $W_{eff}$ of nonplanar devices extends partly upward from the substrate. This requires tight fin pitch but yields higher drive current, with no corresponding penalty in standard-cell or SRAM-cell area.
- **Enhanced mobility:** The finFET has proven to be compatible with the various strain-engineering techniques used to enhance mobility in planar MOS devices. Its vertical fin may enable still other refinements. While the face of the standard silicon wafer lies in the (100) crystal plane, individual fins could be oriented along

other axes, characterized by higher hole mobility $\mu_p$ (or by equal $\mu_n$ and $\mu_p$). Such process refinements might reduce the need to widen *p*-type devices.

- **LP, HP, I/O device types:** The finFET transistor has qualified as a platform SOC technology. It can support a mixed family of devices on one die, including low-power, high-performance, higher (I/O) voltage, SRAM, and analog/RF. As one example, Intel's 22-nm process employed longer (40-nm) finFETs for either ultra-low-power logic with a very low subthreshold current or higher-voltage devices to handle legacy 1.8 or 3.3 V I/Os, voltage regulators, or transceivers [42]. Several representative strengths of the finFET device are summarized below:
- **Quantized width:** The effective finFET transistor width $W_{eff}$ is quantized and can be adjusted higher only by laying out more fins connected in shunt. This complicates layout. Precise device sizing is harder, making it difficult to trade off performance and power. A tight fin pitch is essential for fine-grained width adjustment. This is especially true for sizing and floor planning analog blocks, since device widths are quantized, and the fins must be laid out in parallel lines.
- **Overheating in fin:** Localized overheating can result from high-drive currents flowing through thin fins. The heating can degrade $I_{ON}$. In the long term, it may be a reliability problem. Less of an issue for *bulk* finFETs, whose fins are thermally coupled to the substrate, it may be more of an issue for silicon-on-insulator (SOI) finFETs, fabricated upon a buried-oxide layer. The thermal conductivity of a thin layer of $SiO_2$ can be two orders of magnitude lower than for bulk silicon.
- **Parasitic capacitance:** Parasitic values are process-specific. Because the fin is a vertical ridge, finFETs tend to have less parasitic junction capacitance $C_j$ than planar MOSFETs, with their wide, flat source and drain areas. But a tall fin and gate stack, in close proximity to local interconnect lines, can result in higher gate-to-source/drain parasitic capacitance $C_{gs}$ between the gate and the exposed fins or METAL0 lines forming the common source and drain. In general, the 3-D structure of a finFET complicates the extraction of parasitic *RC* values from device geometry. There are more parasitic elements to compute, per transistor.
- **Parasitic resistance:** High parasitic resistance can arise from the source and drain terminals at both ends of the thin fin, compared to the planar MOSFET. This was evident in Fig. 2.7. High series *R* can impact on-chip RF circuits via lower transconductance $g_m$, which affects amplifier gain or frequency response.

Though accurate long-term predictions are not easy to make in the fast-changing semiconductor industry, the nonplanar finFET device, with its almost wraparound gate terminal and its maximal electrostatic control over the channel, promises to extend Moore's law, avoiding short-channel effects, to at least the 3-nm node [44].

## Glossary

**Analog/RF CMOS** As used in this tutorial, it refers to analog circuits (from low- to radio-frequency and even millimeter wave) on a mostly digital CMOS chip. This presents a challenge for the analog designer, since any CMOS fabrication process is optimized for noisy digital logic, not sensitive analog functions such as low-noise amplifiers or wireless transceivers. Digital processes offer simple speed vs. power trade-offs, while analog circuits may need optimizing for gain, linearity, or noise.

**BSIM SPICE models** The acronym stands for Berkeley short-channel IGFET (insulated gate) model. Older versions, like BSIM3 and BSIM4, model traditional MOSFETs, up to the 22-nm node. Newer versions like BSIM-CMG model nonplanar devices like the finFET. These models stem from the principles of device physics (like the GCA equations used in this tutorial), rather than purely empirical formulas.

**Carrier concentration** In contrast to metallic conductors such as aluminum or copper, semiconductors have two *independent* charge carriers: electrons and holes. The behavior of diodes, transistors, and other semiconductor devices often hinges upon the relative concentrations of holes ($p$) and electrons ($n$), expressed in carriers per cm$^3$. In pure silicon, they are thermally generated in pairs. In this *intrinsic* state, $n$ and $p$ remain equal. Their product is $np = n_i^2 \approx 10^{10}$ cm$^{-3}$ at room temperature.In a $p$-type substrate, however, the substitution of acceptor atoms (like boron) into the silicon lattice results in an excess of *holes*. Then $p_p >> n_p$, indicating an excess of majority-carrier holes over minority-carrier electrons. The product $np = n_i^2$ will remain constant under thermal equilibrium (no applied bias or other perturbations).

**CD** An acronym familiar to fab engineers, denoting any *critical dimension* of a feature on a wafer. CDs can be measured in the fab by a CD-SEM. For example, an Intel engineer might plot subthreshold swing as a function of the measured width of a poly line crossing a fin, in nm. This CD is a routine measure of channel length $L$.

**Drain-induced barrier lowering (DIBL)** When the drain bias of a short-channel device increases from low to saturated values, threshold roll-off is *aggravated.* (The shaded triangles in Fig. 2.1 indicated how roll-off occurs at *low* drain bias. DIBL occurs at higher $V_{DS}$ values.) The drain bias affects the potential distribution, lowering barriers to subthreshold conduction. This SCE is measured in mV per volt.

**Inversion** In an $n$-type MOS transistor at the onset of inversion, the concentration of minority electrons flooding into the channel *just equals* that of holes deep within the $p$-type substrate. Thus, $n_p(0) = p_p(\infty)$, where coordinate 0 refers to the silicon surface. The relative concentration of minority and majority carriers thus becomes *inverted.* Above this threshold, $n_p(0)$ rises *exponentially* with surface potential $\Phi_s$.

**Permittivity $\varepsilon_0$** Intuitively, the electrical permittivity $\varepsilon_0$ of free space is equal to the capacitance of a cube-shaped parallel-plate capacitor, filled with a vacuum, of 1 cm thickness and 1 $cm^2$ area. A fundamental constant, it equals $8.85 \times 10^{-14}$ F/cm.

**PTM SPICE models** The acronym is for predictive technology model, which evolved from predictive modeling efforts at Berkeley, and is now advanced by the Nanoscale Integration and Modeling Group at the Arizona State University. A goal of this group is *predictive* modeling, applicable to the next generation of technology—not only finFET devices down to 7 nm but even to future carbon nanotube devices.

**Silicon on insulator (SOI)** A promising alternative to bulk finFET technology, in which transistors are fabricated directly on a silicon wafer. In SOI technology, an insulating oxide layer (the buried oxide, or BOX) is implanted below the surface of the entire wafer. FinFET transistors are fabricated into the thin layer of silicon left on top. Due to the BOX, there is little or no parasitic capacitance to the substrate. Another advantage of SOI wafers is that no *well* is needed around *p*-type devices. The trade-off is a higher SOI base wafer cost—as much as 4× higher than bulk.

**Subthreshold swing (SS)** Subthreshold current is one of the worst SCEs. It makes a short-channel transistor *too leaky* to turn completely off. Since this current falls off exponentially as $V_{GS}$ drops below $V_T$, it is ordinarily plotted on a log scale. The slope of this log-linear plot is then $\partial \log(I_{DS})/\partial V_{GS}$. In the lab, it is more convenient to measure the *reciprocal* of this slope, in mV per decade of current, often called subthreshold *swing*. It has a theoretical minimum of 60 mV/decade—which Intel nearly met with its 22-nm finFET. Devices which swing higher are more leaky.

**Verilog-A** An addition to the Verilog digital hardware description language was able to model *analog* circuits as well. It includes an `analog` code block, as well as the usual `always`. Since BSIM-CMG is written in Verilog-A, it can use classic `case` syntax, like `case (GEOMOD)`, to check a parameter and set values accordingly. It can also use analog operators like `ddt (q)` to differentiate charge, yielding current.

## References


1. Moore G (1965) Cramming more components onto integrated circuits. Electronics 38:114
2. Thackray A (2015) Moore's law. Basic Books, New York, p 243, 253–262
3. Dunga MV, Lin C-H, Niknejad AM, Hu C (2008) In: Colinge J-P (ed) FinFETs and other multi-gate transistors. Springer, New York, Table 3.1
4. James D Intel Ivy Bridge unveiled (*Chipworks*), djames@chipworks.com
5. Gargini P (Intel Fellow) (2003) In: NanoManufacturing symposium (2003), p 19, from: http://web.mit.edu/nanosymposium/www/Presentations/Gargini2Part1.pdf
6. Xiong W (2008) In: Colinge J-P (ed) FinFETs and other multi-gate transistors. Springer, New York, Section 2.2.6

7. Halliday D, Resnick R, Walker J (1993) Fundamentals of physics, 4 edn. Wiley, New York, Section 45–4
8. Kim DM et al (2014) In: Kim DM, Jeong Y-H (eds) Nanowire FETs: principles and applications. Springer, New York, Section 4.1
9. Colinge J-P (ed) (2008) In: FinFETs and other multi-gate transistors, Springer, New York, Section 1.5.2.1
10. Colinge J-P, Greer JC (2016) Nanowire transistors: physics of devices and materials in one dimension. Cambridge University Press, Cambridge, Equation 2.20
11. Routenberg D Fabrication and characterization of silicon nanowire field effect sensors. PhD Thesis, Yale University, Section 2.1
12. US Patent 2017/0133509 A1, FinFET, Taiwan Semiconductor Manufacturing Co. (2017) Section [0036]
13. Craig Casey H Jr (1999) Devices for integrated circuits. Wiley, New York, Figure 8–6
14. Craig Casey H Jr (1999) Devices for integrated circuits. Wiley, New York, p 291
15. Mead C, Conway L (1980) Introduction to VLSI systems. Addison-Wesley, Reading, Equation (1–1)
16. Pierret R (1990) Field effect devices. Modular series on solid-state devices, vol. IV. Addison-Wesley, Reading, Figure 3.8
17. Sze SM (2002) Semiconductor devices: physics and technology, 2 edn. Wiley, Hoboken, Section 6.2.1
18. Pierret R (1990) Field effect devices. Modular series on solid-state devices, vol. IV. Addison-Wesley, Reading, p 74
19. Paydavosi N et al BSIM-SPICE models enable FinFET and UTB IC designs (*IEEE Access*, Vol.1, 2013), Equation (1)
20. Chauhan YS et al (2015) FinFET modeling for IC simulation and design: using the BSIM-CMG standard. Elsevier, Amsterdam, Equation 3.12
21. See for example: Moroz V FinFET structure design & variability analysis enabled by TCAD (*EE Times*, 8 Oct. 2012) Figure 3
22. Chauhan YS et al (2015) FinFET modeling for IC simulation and design: using the BSIM-CMG standard. Elsevier, Amsterdam, p 71
23. Paydavosi N et al BSIM-SPICE models enable FinFET and UTB IC designs (*IEEE Access*, Vol.1, 2013) Figure 11
24. Farkhani H, Peiravi A, Kargaard JM, Moradi F (2014) In: Comparative study of FinFETs versus 22nm bulk CMOS technologies: SRAM design perspective. IEEE SOCC, Figure 2
25. Routenberg D Fabrication and characterization of silicon nanowire field effect sensors. PhD Thesis, Yale University, Equation 2.42
26. Colinge J-P (ed) (2008) In: FinFETs and other multi-gate transistors. Springer, New York, Section 1.5.1.3
27. Ferain I, Colinge CA, Colinge J-P Multigate transistors as the future of classical MOSFETs. NATURE 479, 17 Nov. 2011, Figure 5a
28. BSIM-CMG 110.0.0 technical manual, multi-gate MOSFET compact model. EECS, University of California, Berkeley, 2015, Section 3.15.2.2
29. Pierret R (1990) Field effect devices. Modular series on solid-state devices, vol. IV. Addison-Wesley, Reading, Figure 1.7
30. Farkhani H, Peiravi A, Kargaard JM, Moradi F (2014) In: Comparative study of FinFETs versus 22nm bulk CMOS technologies: SRAM design perspective. IEEE SOCC, Section II.A
31. Lapedus M Re-engineering the FinFET. Semiconductor Engineering, Nov. 2014, http://semiengineering.com/re-engineering-the-finfet
32. Colinge J-P (ed) (2008) FinFETs and other multi-gate transistors. Springer, New York, p 25
33. Xiong W (2008) In: Colinge J-P (ed) FinFETs and other multi-gate transistors, Springer, New York, Section 2.2.3
34. Pinto R Analog designer, Oracle, personal discussion
35. Payne D FinFET modeling and extraction at 16-nm. SemiWiki.com, Dec. 2012

36. Yan RH, Ourmazd A, Lee KF (1992) In: Scaling the Si MOSFET: from bulk to SO1 to bulk. IEEE TRANS. ELECTRON DEVICES, vol. 39, Equation (11)
37. Colinge J-P (ed) (2008) In: FinFETs and other multi-gate transistors. Springer, New York, Section 1.5.1.1
38. Colinge J-P (2004) Multiple-gate SOI MOSFETs. Solid-State Electronics 48, Equation (9)
39. Collaert AN (ed) (2013) In: CMOS nanoelectronics: innovative devices, architectures, and applications. Pan Stanford Publishing, Singapore, Table 1.1
40. Colinge J-P (ed) (2008) In: FinFETs and other multi-gate transistors. Springer, New York, Table 1.2
41. Ferain I, Colinge CA, Colinge J-P Multigate transistors as the future of classical MOSFETs. NATURE, 479, 17 Nov. 2011, Figure 8
42. Jan C-H et al A 22-nm SOC platform technology featuring 3-D tri-gate and high-k/metal gate. Intel Corporation, Hillsboro
43. Natarajan S et al (2014) A 14-nm logic technology featuring 2-G FinFET transistors. IEDM, Intel Corporation, Hillsboro
44. Ferain I, Colinge CA, Colinge J-P Multigate transistors as the future of classical MOSFETs. NATURE, 479, 17 Nov. 2011

# Chapter 3
# Fundamentals of Metal-Oxide Resistive Random Access Memory (RRAM)

**David C. Gilmer and Gennadi Bersuker**

**Abstract** Detailed operational and switching characteristics for metal-oxide resistive random access memory (RRAM) are presented, along with materials/vacancy engineering ramifications for the switching operations. The reported experimental data is consistent with a physical picture of the RRAM switching as caused by oxidation/reduction processes in a conductive filament formed through the metal-oxide material. The oxygen interstitial ion and vacancy profile resulting from the filament formation process establishes the initial structural properties controlling subsequent reset/set resistance switching operations. The microscopic description of these processes links the device electrical and material characteristics allowing for optimized material compositional profiles to improve device performance characteristics.

## 3.1 Introduction

Continued progress in currently used memory systems (high-density memory cell arrays, dynamic random access memories (DRAM), NAND, storage class memory (SCM), advanced embedded type applications, etc.) depend on continued advances toward low operation currents and voltages while increasing density and speed. In addition, the rising requirements for reducing power consumption in mobile applications with convenience of use have increased the efforts and focus toward technology development of new nonvolatile memories [1]. In this respect, the resistance switching random access memory (RRAM) technology presents an attractive option due to demonstrated potential for low-complexity/high-density/low-cost nonvolatile

D. C. Gilmer (✉)
Nantero, Inc., Austin, TX, USA
e-mail: david.gilmer@nantero.com

G. Bersuker
Aerospace Corporation, El Segundo, CA, USA



S. M. Goodnick et al. (eds.), *Semiconductor Nanotechnology*, Nanostructure Science and Technology, https://doi.org/10.1007/978-3-319-91896-9_3

memory cells with high-speed/low-energy operations and prospective ability to satisfy the requirements of various incumbent advanced scaled memory systems currently in use [2, 3]. A common characteristic within the large family of the metal-oxide-based resistance switching memory schemes is that their operating mechanisms involve either rearranging the atomic structure of the dielectric material (rendering it conductive) or movement of atoms in the dielectric (resulting in the formation of a conductive path), as opposed to the current incumbent memory technologies based on the electron movement and storage. This overview is focused on the transition metal-oxide filament-based type of RRAM (with emphasis on hafnium oxides [HfOx]), which offers promising opportunities for a variety of nonvolatile memory applications.

Metal-oxide-based filament-type RRAM utilizes the repeatable formation and rupture of a localized conductive filament through the metal oxide and has the unique attribute of area-independent resistance. This filament-based mechanism suggests an ultimate scaling advantage that is only limited to the active filament size, which potentially can be only a few nm. The mechanisms of filament-type resistive switching depend on the materials (dielectric and metal electrodes) employed in the fabrication of the memory cell and can involve more than one type of conduction mode. Filament-based metal-oxide RRAM schemes have been implemented with a variety of transition metal oxides (such as $HfO_2$, $ZrO_2$, $Ta_2O_5$, $TiO_2$) which have recently received considerable attention due to their demonstrated nanosecond, low-power (<pJ) switching with high (~trillion cycle) endurance and retention of >10 years at 200C using fabrication-friendly binary oxide dielectrics and metal electrodes [4–12]. Among the metal-oxide dielectrics in this group, $HfO_2$ has recently become mainstream in advanced transistor gate stack applications and has also become one of the stronger candidates for RRAM [4, 6, 9, 10, 13, 14]. One of the advantages in using $HfO_2$ for metal-oxide filament-based RRAM is the extensive research performed during ~1997–2007 in developing gate stacks toward manufacturability and productization of advanced logic using metal gate and $HfO_2$ as the "high-K" gate insulator, which enabled the detailed understanding of $HfO_2$ resistivity, conductivity, defects, and breakdown mechanisms [15–18]. In addition, $HfO_2$ has one of the stronger oxygen affinities among the transition metals, and thus, thermodynamically speaking, it is relatively stable and compatible with most of the commonly used "fab-friendly" metallic electrodes such as TiN, TaN, W, etc. [19–24].

This review focuses primarily on the $HfO_2$-based RRAM system, discussing its operation and optimization of device characteristics through materials (and defect/vacancy) engineering along with a discussion of the possible mechanisms involved in the observed repeatable resistance switching. Although some of the processes contributing to resistive switching in the $HfO_2$-based RRAM may be active in other metal-oxide RRAM systems, due to the material-specific nature of RRAM characteristics (such as relative oxygen affinities, valance states, and atom diffusivities), any conclusions or comparisons to other metal-oxide material systems must be made with caution.

## 3.2 Operational Characteristics for $HfO_2$-Based RRAM

Typically, RRAM cells employ a metal-insulator-metal (MIM) structure, which is often referred to as a 1-resistor (or 1R) configuration. The resistance change behavior in bipolar operation for filament-based RRAM switching, as observed for $HfO_2$, is shown in Fig. 3.1a where an ohmic low resistance "on" state (LRS) or a non-ohmic high resistance "off" state (HRS) can be related to the formed conductive filament (CF) or its rupture, respectively (Fig. 3.1b). To describe the material changes in the dielectric associated with this type of resistive switching, one should start with the CF formation, where a microscopic description of the CF features and metal-oxide defect/vacancy engineering that enable memory operations have recently been proposed [25, 26]. The CF forming process in $HfO_2$ can be discussed in terms of dielectric breakdown [4], where there is an abrupt formation of a localized region between the electrodes, within which the dielectric composition becomes more oxygen-deficient (or metal-rich), rendering this region conductive (the CF formation). The associated abrupt conductance change for this CF formation is observed by applying a voltage across the RRAM dielectric in the fresh ("virgin") state, Fig. 3.2. The typical bipolar *dc-iv* switching operation are noted on the graph: (1) forming [the initial dielectric breakdown event leading to a CF with the voltage of the abrupt breakdown (forming) termed $V_F$]; (2) RESET operation [at opposite voltage polarity (than forming or set) for the bipolar operation [*as opposed to forming/reset/set all at same polarity in unipolar switching*]], where the device changes from the LRS to the HRS state due to rupture of the CF; and (3) SET operation [at opposite voltage polarity vs reset and similar polarity as forming], where the device changes from the HRS to the LRS state due to reformation of a continuous CF between the two electrodes.

The typical *dc-iv* switching operation parameters and device response in bipolar-operated RRAM devices are shown in Fig. 3.3. Similar well-behaved devices can repeatedly be switched between the LRS and HRS states for billions of cycles, with the corresponding switching operations being reset (LRS→HRS) and set (HRS→LRS) [6, 7, 9, 14, 26].

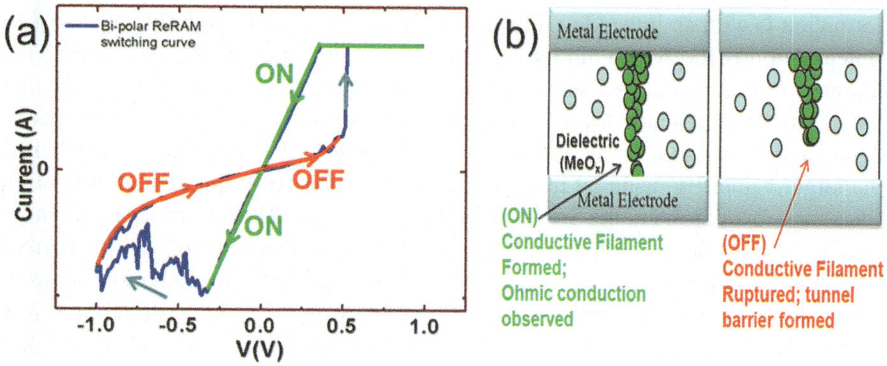


**Fig. 3.1** (**a**) Characteristic dc properties of the on (LRS) and off (HRS) states and (**b**) related physical picture for filament-based RRAM

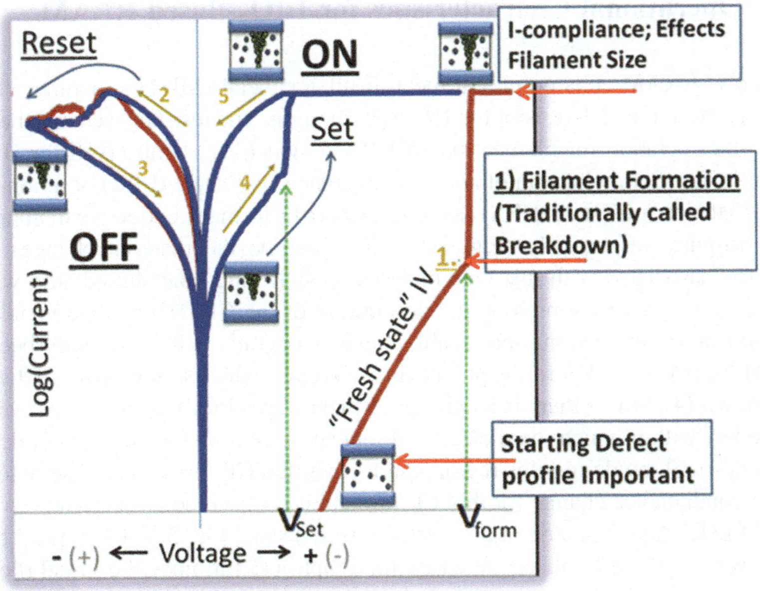


**Fig. 3.2** Characteristic dc operation of $HfO_2$-based RRAM devices highlighting (1) forming [a dielectric breakdown event leading to a conductive filament (CF)]; (2) RESET operation, where the device changes from the LRS to the HRS state due to rupture of the CF; and (3) SET operation, where the device changes from the HRS to the LRS state due to reformation of the CF path

### *3.2.1 Current Compliance and 1T1R*

As seen in Fig. 3.2, a compliance is used to control the current during the CF (dielectric breakdown). If the current is not limited during this step (or during set), the CF will be uncontrolled and too large which would negatively affect the reset and subsequent switching operations. The effective way to control the current during CF formation is with an imbedded (integrated) transistor in series with the memory cell (integrated 1T1R), which limits the current passing through the cell during the filament forming step (and set operations) thus preventing current from exceeding the desirable maximum level (current overshoot issue) [6, 7, 27–29]; however, conductivity overshoot may still occur [4] resulting in CF resistance being below the target value. By using an integrated transistor to control the filament formation (dielectric breakdown), the intrinsic properties of the RRAM devices can more clearly be revealed. The parasitic capacitance, which may also lead to current overshoot, can be limited, in general, by using small area drain, bottom and top electrodes, small and/or mesh-patterned probe pads, and thick isolation layers surrounding the RRAM cell. Using the special device structure shown in Fig. 3.4, which can allow for testing the same RRAM element in various different configurations of integrated 1T1R, or with external parametric analyzer, or even with an

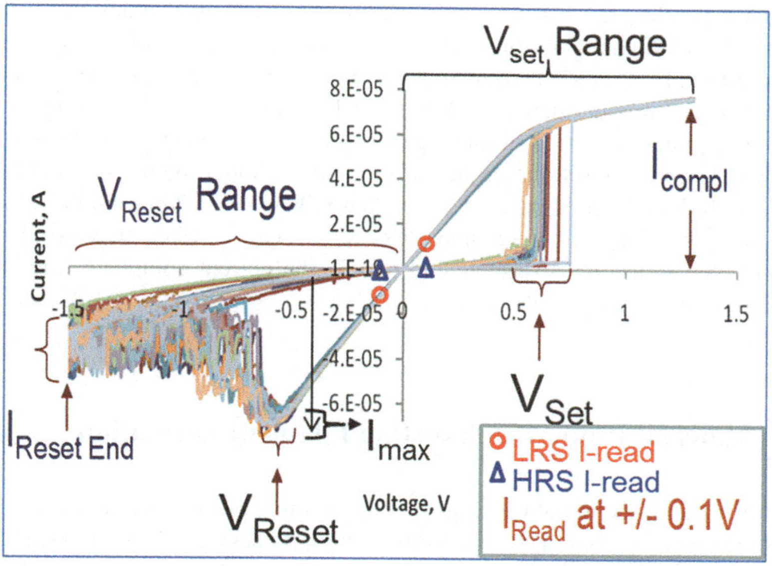


**Fig. 3.3** A summary of the typical dc switching operation parameter terms for the bipolar-operated RRAM devices: Vset, V reset, Vrest range, Icomp, Imax, LRS I-Read, and HRS I-read. Vreset is defined by the onset of current reduction trend, Vreset range is the maximum voltage magnitude applied during the reset operation. Note that in DC operations, "Vreset range" is usually referred to as the reset voltage

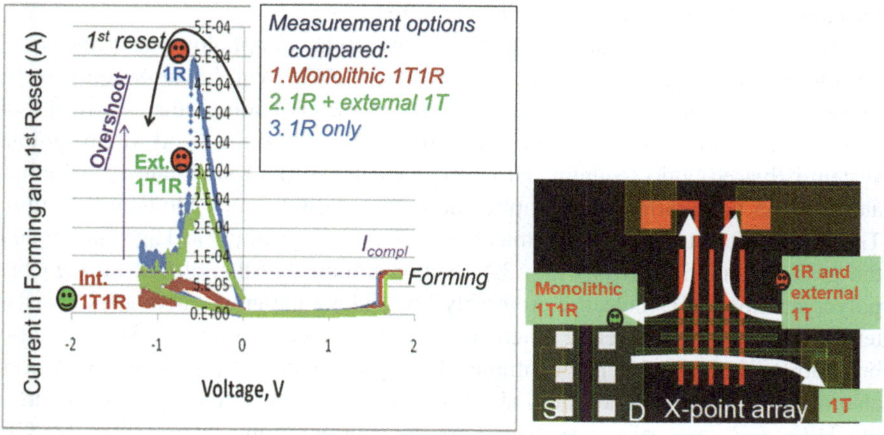


**Fig. 3.4** Impact of parasitics on reset current. DC sweeps of forming and first reset on 100nmx100nm crossbar *TiN/Ti/HfOx/TiN* RRAM stack for the device configurations shown having compliance limited by integrated transistor (Int. 1T1R), externally connected transistor (Ext. 1T1R), or externally connected parametric analyzer (1R). Only the integrated transistor results in no current overshoot during first reset above the forming compliance value applied during CF formation

external transistor, to limit operation current, clearly demonstrates that only the integrated 1T1R configuration leads to elimination of the parasitic capacitance (as manifested in current overshoot of Imax above the compliance current [Icomp]) [7]. However, there are additional CF forming operation techniques, such as "multiforming" [30] and ultrafast pulse forming [4], which have demonstrated improved current control during the CF formation without the use of an integrated transistor. It should also be noted that there have likely been a large number of metal-oxide samples discarded upon poor RRAM switching, when they would have demonstrated stellar RRAM switching if only the current compliance during CF formation was properly controlled.

## 3.3 Material Properties Assisting Filament Formation

The properties of the initial CF created in the forming process determine the device switching characteristics. The successful forming process involves applying a sufficient bias across the metal-oxide dielectric resulting in the formation of a CF with ohmic- or near-ohmic-type conduction (as observed from the near linear I-V dependency after CF formation and increasing resistance with temperature). For the $HfO_2$ dielectric, these metallic characteristics indicate that the filament is represented by a Hf-rich/oxygen-deficient region in the dielectric. The filament formation is thus associated with the process of oxygen expulsion from this dielectric region, which is the focus of the following discussion. To start, an understanding of the stoichiometric and morphological properties of the $HfO_2$ dielectric film which are assisting formation of the conductive filament is required.

To assist the understanding toward connecting morphology and electrical response, conductive atomic force microscopy (C-AFM) can be an effective technique [31–33], allowing nanometer-resolved characterization of the electrical and topographical properties of oxide dielectrics [34]. When the C-AFM tip-sample system is biased, a tip scanning across the dielectric surface simultaneously generates data on both surface topography and current flow through the dielectric film. The C-AFM results show a good match between the maps of the leakage current and the grain boundaries (GBs) (Fig. 3.5) demonstrating that the current through the polycrystalline $HfO_2$ dielectric preferably flows along the grain boundaries, consistent with scanning tunneling microscopy (STM) measurements [35]. Dielectric breakdown, induced by the continuously applied voltage, preferentially occurs also at the GB sites. In addition, ab initio calculations were employed to identify the $HfO_2$ GB properties responsible for the current flow, and despite the GB stability, segregation of vacancies to the boundary is thermodynamically favorable [36].

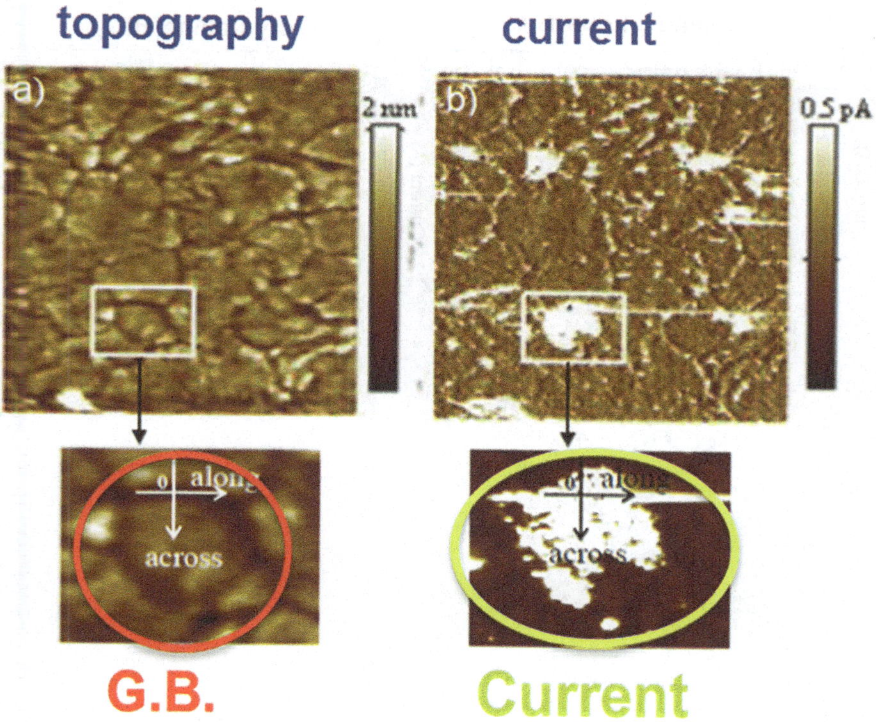


**Fig. 3.5** Topographical (**a**) and current (**b**) images of a 5 nm $HfO_2$ film obtained with C-AFM. Multiple scans resulted in breakdown spots formed at the sites of the grain boundaries

## *3.3.1 The Role of Grain Boundaries in Electron Transport Through Hafnia*

Electrical transport along the GBs in polycrystalline $HfO_2$ can be successfully described by considering the multi-phonon trap-assisted tunneling (TAT) process [26, 37] via vacancy sites. The leakage current through a TiN/$HfO_2$/TiN capacitor was simulated, considering both direct tunneling (DT) and TAT components of the gate current [36, 37], and successfully reproduced the current-voltage (I-V) temperature dependency in the polycrystalline $HfO_2$ [38]. The TAT current is the dominant component in dielectric films thicker than about 4 nm and is calculated by taking into account contributions from both the single-trap and multi-trap conductive paths by defects randomly placed along the GBs. The extracted values for the energy of the contributing traps and their relaxation energy match those calculated for the singly positively charged oxygen vacancy ($V^+$) at the GBs, indicating that the TAT current is associated with the ($V^+$+e $\rightarrow V^0 \rightarrow$ $V^+$+e) process [26]. The doubly positively charged oxygen vacancies ($V^{2+}$) exhibit the lowest diffusion activation energy (about 0.7 eV), and their segregation at the GBs leads to the formation of a conductive sub-band within the dielectric energy gap. By capturing injected

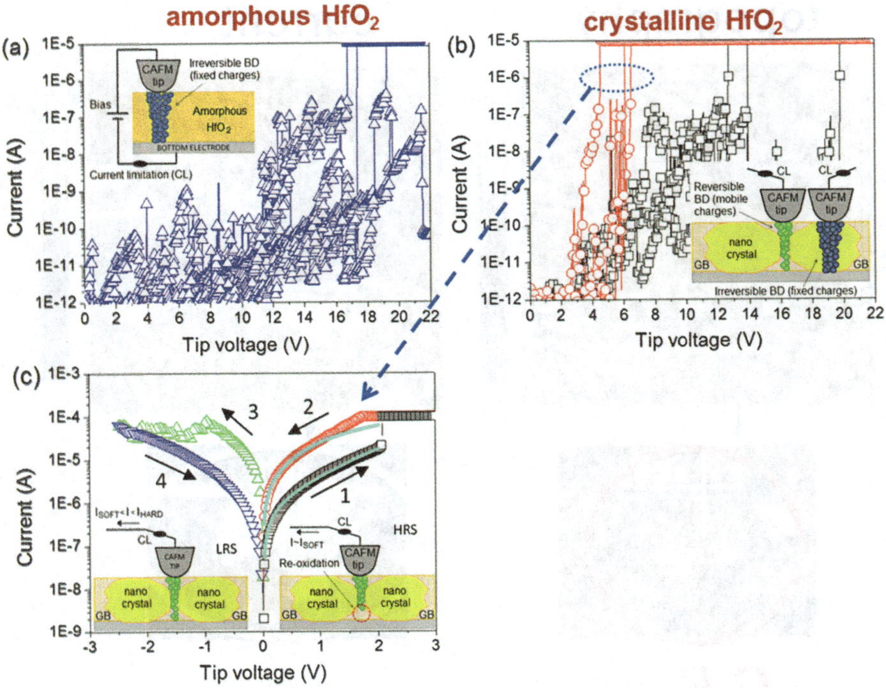


**Fig. 3.6** Forming process at different random locations in (**a**) sample with non-annealed amorphous 3 nm $HfO_2$ and (**b**) annealed polycrystalline 3 nm $HfO_2$ where two different I-V patterns associated with low-voltage and high-voltage forming can be distinguished. (**c**) An example of switching observed at the low-voltage forming site in (**b**). The schematics indicate the probing location and measurement conditions

electrons when the forming voltage is applied, these vacancies are converted to a $V^+$ state, which is shown to support the TAT current by quickly capturing/releasing subsequently injected electrons [26]. The current flowing along the GBs thus facilitates a dielectric breakdown along one of the GB paths, which results in the preferential formation of the conductive filaments at the GB sites in polycrystalline metal oxides such as $HfO_2$.

To evaluate the effect of GBs on the filament formation, the forming/switching operations were performed in nanoscale by using the tip of a C-AFM probe as a top electrode of a MIM device (the biased TiN substrate constitutes the bottom electrode) (Fig. 3.6). The higher conductivity at the GBs in the polycrystalline film is accompanied by lower forming (breakdown) voltages. These lower forming voltage sites at GBs exhibit repeatable switching between high and low resistance states, while no resistive switching is observed when the forming is induced on the grain sites (as is also the case for an amorphous stoichiometric $HfO_2$), which exhibit significantly higher forming voltage values. These results indicate that resistive switching benefits from higher oxygen deficiency of the dielectric GBs making them easier to convert to a CF state at lower forming voltage [26].

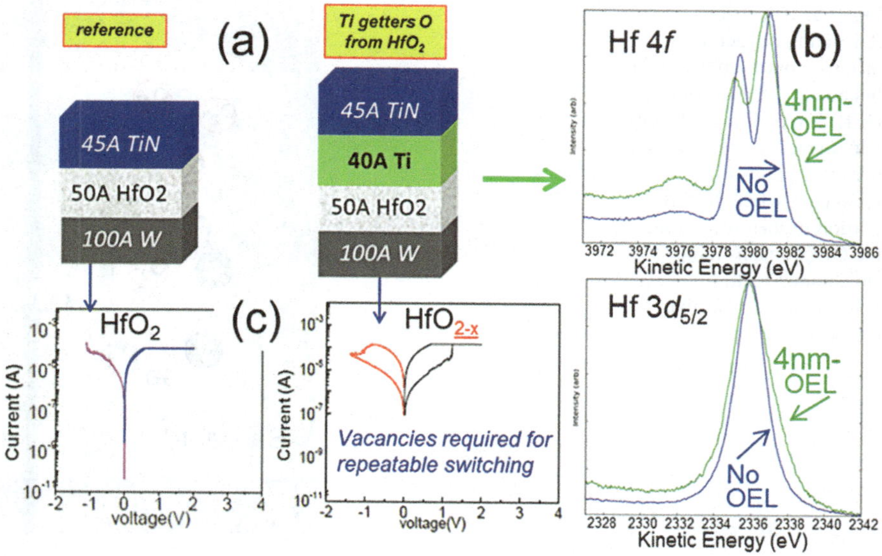


**Fig. 3.7** Evidence of $HfO_2$ reduction by Ti-OEL; schematic of the stacks measured by the synchrotron XPS [at the National Synchrotron Light Source (NSLS) at (BNL)] (**a**), Hf 4f and $3d_{5/2}$ spectra (**b**), and switching characteristics of the corresponding devices (**c**)

It is also observed in MIM devices (such as TiN/$HfO_2$/TiN) that although perfect or near perfect stoichiometric $HfO_2$ does not readily demonstrate repeatable resistive switching after CF formation, the nonstoichiometric $HfO_{2-x}$, with "some" oxygen deficiency (density of vacancies), does exhibit resistive switching after forming. This is highlighted in Fig. 3.7 where an overlaying thin metal film can be used to scavenge or "getter" oxygen from stoichiometric $HfO_2$ dielectric leading to the oxygen deficiency as shown by the synchrotron XPS data. The sub-stoichiometric $HfO_{2-x}$ film, with certain oxygen vacancies densities, demonstrates much better switching characteristics than the stoichiometric $HfO_2$ film. Thus, to create oxygen vacancies facilitating the CF formation, the gettering metal for extracting oxygens from the metal oxide is deposited over the metal-oxide film. Such gettering metal layers are usually early transitional metals (i.e., Ti, Zr, Hf, etc.) and often termed oxygen exchange layers (OEL) in RRAM stacks [26]. In addition, physical observations combined with electrical data have implicated $HfO_2$ GBs as susceptible toward reduction even when not favored from bulk thermodynamic considerations [39]. It is also noted that a typical thickness of a gettering metal in this application is 20A-100A and their gettering power grows with their thickness. Capping the gettering metal in a process that does not contain reactive nitrogen (which can reduce the metal gettering ability by nitriding it), like with in situ PVD tungsten after the gettering metal deposition, also increases the oxygen gettering ability of the OEL used.

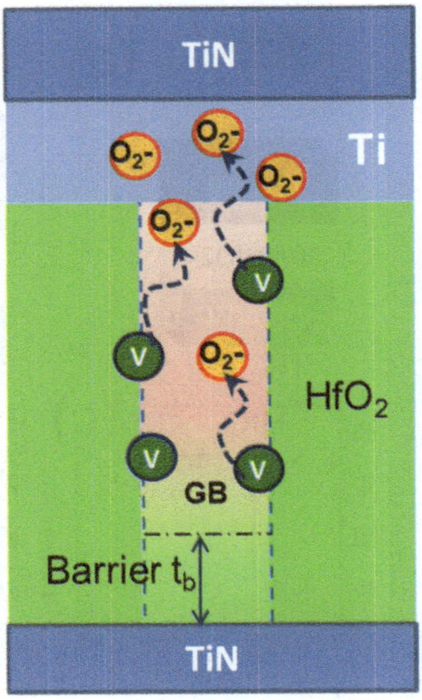


**Fig. 3.8** Schematic illustrating oxygen ions diffusing out from the grain boundary to the overlaying OEL. The portion of the GB with higher oxygen vacancy density exhibits low resistance, while the GB portion further away from the OEL remains more stoichiometric, thus constituting an effective dielectric barrier

Using the above-described stack, the $HfO_2$ GB portion next to a gettering metal film is less resistive (due to higher concentration of the oxygen vacancies there), while a "more" stoichiometric or less "gettered" section of the GB away from gettering metal is a "better" more resistive dielectric due to less oxygen vacancies there (Fig. 3.8). Such a random process of oxygen gettering is expected to proceed differently in each GB due to variability of the GB characteristics (size, width, length, orientation, structure, etc.) resulting in a distribution of thicknesses of the "more" dielectric behaving barrier in the GB away from the gettering metal which would gate the current in the various GBs of each given device and, subsequently, a distribution of the magnitudes of leakage currents through these various GBs. The leakiest GB path is then accordingly expected to have the thinnest "dielectric" barrier since the electron transport through the barrier controls the overall current and the GB with the thinnest barrier should be the first one, out of all the other GB barriers with less conductive paths, to be broken by the applied voltage during CF forming. Thus, a single structural feature – the thinnest GB barrier in a given device – is responsible for both the lowest breakdown (CF forming) voltage magnitude and the highest leakage current (which is proportional to the total current through the device) and explains their correlation [26, 40]. Consequently, the primary beneficial effect of oxygen gettering is the creation of relatively highly conductive GB current paths (with higher oxygen deficiency toward the gettering metal), which can exhibit low forming voltages leading to a controlled CF formation that subsequently can enable the repeatable resistive switching phenomenon (i.e., continued rupture and forming of the CF).

## 3.4 Vacancy Asymmetry Engineering and the Metal-Oxide Filament-Based Switching Model

As described above, oxygen vacancies in sub-stoichiometric films may accumulate along the GBs in polycrystalline films [41], thus forming "preferred" conductive paths through the dielectric. The presence of the conductive paths along GBs assists with lowering the filament forming voltage allowing for more controlled and smaller conductive filament (CF) formation. Another important phenomenon of the CF is its nonuniform cross section resulting from the formation physics, giving it a cone-like shape (see Fig. 3.9a) [25, 26]. The cone-like CF shape provides certain advantages for switching operations, with bipolar reset described as oxidation of the narrower end of the CF tip to form a thin dielectric barrier near the positively biased reset-anode (which for bipolar operation is also near the forming cathode) leading to the high resistance state (HRS) (see Fig. 3.9b). The subsequent set operation (from HRS to the low resistance state [LRS]) is then the breakdown of this thin dielectric barrier thus reforming the CF. For a device to demonstrate well-behaved RRAM switching, the forming and switching biasing conditions should be selected according to the filament composition profile to achieve favorable distribution of the oxygen ions, which subsequently contribute to reoxidation of the filament tip during reset.

The *dc-iv* reset process in an HfOx-based RRAM device is illustrated in Fig. 3.10 [26, 41, 42]. Changes in the LRS during increasing reset bias are first observed by deviation off the linear I-V curve as could be expected for an ohmic LRS due to

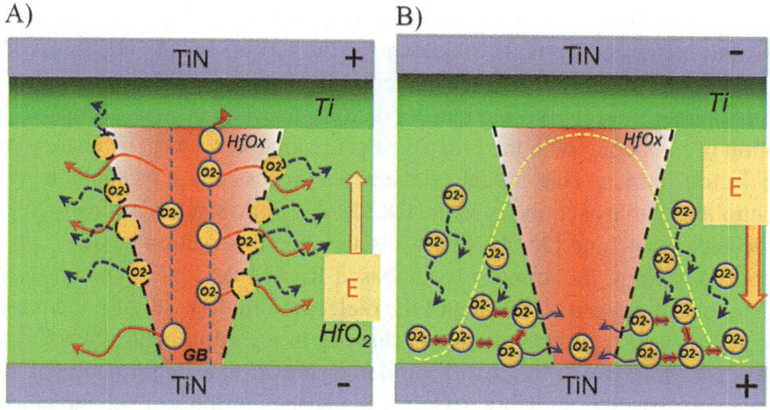


**Fig. 3.9** (**a**) The filament formation process. The filament growth is initiated at GB (vertical broken lines) where the oxygen bonding is weaker than in the bulk oxide and proceeds via the process of the O(2-) ions release from the filament outer surface and moving with the electric field and also away from the filament due to concentration gradients. *These O-ions do not move too far as they are trapped by charges associated with grain boundaries and thus they are then readily available for CF oxidation in reset [26]. (**b**) For the reset process, the broken lines and broken bell-shape curve outline the filament and temperature radial profile, respectively. The nearby "available" O-ions diffuse following the electric field and density gradient and reoxidize a portion of the filament next to the positive anode electrode forming a thin HRS dielectric barrier

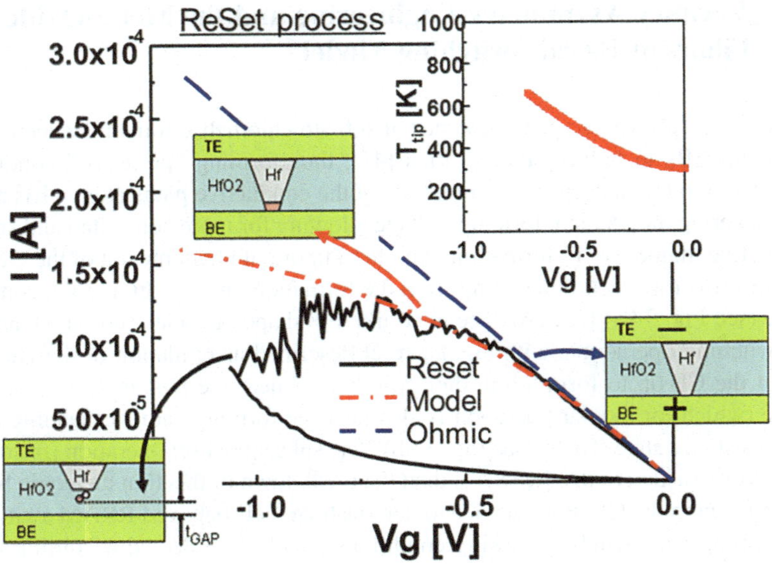


**Fig. 3.10** RESET IV sweep with schematics of the physical process. Initially most voltage drops across the narrow conductive filament tip. The current deviates from ohmic initially due to a resistance increase from increasing filament temperature. Reset then occurs when the temperature is sufficient for oxidation and oxygen is available (i.e., diffusion from nearby "available" oxygen)

temperature increase of the CF (Fig. 3.10 inset shows the simulated temperature increase with increasing bias). Eventually upon continued bias and sufficient heat (due to the increasing current), oxidation of the CF tip begins, as observed by a "noisy" decrease in current with continued voltage increase. This fluctuating resistance magnitude reveals the random oxidation process near the CF tip where CF oxidation is competing with the process of generating new oxygen vacancy pairs in the dielectric barrier. This model of the reset process suggests that the vacancy concentration profile asymmetry in the RRAM dielectric stack is critically important for repeatable switching [26, 42–45] as discussed in this section below.

Robust switching in the metal-oxide-based RRAM devices is known to require some amount of oxygen deficiency of the dielectric film [26, 28, 45]; the reason for this requirement was discussed above primarily in regard to controlling CF formation at lower voltages. It has also been demonstrated that performance of RRAM devices is improved by utilizing asymmetric vacancy profiles of sub-stoichiometric metal-oxide films. These asymmetric vacancy profiles are typically formed using reactive PVD [11, 26, 28, 46], or with depositing a dielectric or metal and applying a subsequent posttreatment reduction or oxidation anneal [47–49], or by depositing reactive gettering-type metals (like described above and often termed oxygen exchange layers (OEL) in RRAM stacks) on to the metal oxide [9, 10, 14, 50, 51]. Many of the various metal-oxide-RRAM reports in the literature also consistently show bipolar operations with the reset bias applied with the positively biased reset-anode against the more oxygen-rich or more stoichiometric part of the engineered dielectric stacks [9, 10, 27, 45, 46, 48–50]. The reason the bipolar operation in metal-oxide filament-type RRAM consistently have this polarity preference characteristic is related to the metal-oxide asymmetric defect/vacancy profile, the CF asymmetry, and the switching mechanism as further described below. Having fewer defects/vacancies near the "reset-anode" allows for a more effective ability toward reoxidation of the narrow conductive filament tip in this area. For the case using an opposite bipolar operational polarity to switch the RRAM device (i.e., with the positively biased reset-anode against the more defects/vacancy-rich or less stoichiometric part of the engineered dielectric stack), too many nearby defects/vacancies and too few "available" oxygen atoms prevent effective oxidation of the CF tip [26, 45].

## 3.5 Switching in Asymmetric Vacancy Engineered RRAM

To investigate the preferred bipolar operational biasing and vacancy asymmetry relationship, various sub-stoichiometric $HfO_x$ films with asymmetric oxygen vacancy profiles were formed [26, 45]. An OEL of titanium was deposited on top $HfO_2$ and subsequently capped in situ with TiN creating an asymmetric defect/vacancy profile with more defects/vacancies near the OEL side of the stack due to preferential gettering of oxygen from the $HfO_2$ grains and grain boundaries nearer the OEL metal. Figure 3.11a, b presents the switching behavior for the two possible bipolar biasing schemes: (1) Fig. 3.11a, with CF formed in a bipolar biasing scheme that has the positively biased reset-anode opposite the OEL (or more defects/vacancies) side, and (2) Fig. 3.11b, with CF formed in a bipolar biasing scheme that has the positively biased reset-anode at the OEL (or more defects/vacancies)

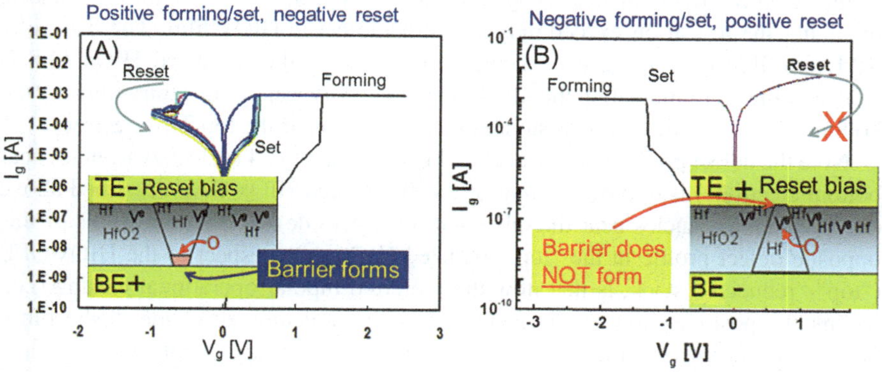


**Fig. 3.11** Results for applying the two possible bipolar operational biasing schemes. When operational biasing has the reset-anode near the more oxygen-rich or stoichiometric side, a dielectric barrier forms successfully achieving reset (**a**). However, having oxygen-poor or higher vacancy/defect concentration near the reset-anode prevents a dielectric barrier from forming, and no reset is observed (**b**)

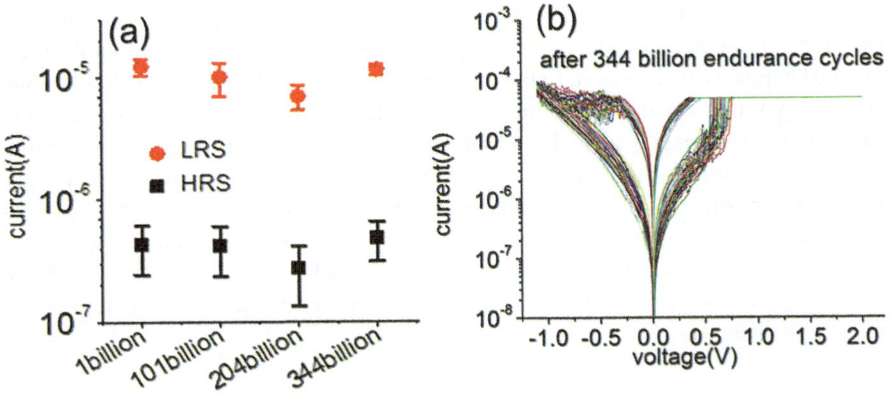


**Fig. 3.12** Endurance results for similar RRAM stack as used for Figs. 3.10 and 3.11 (TiN[$HfO_x$\Ti-OEL]TiN). Averaged LRS and HRS reads from 50 dc cycle checks at 1, 101, 204, and 344 billion ac endurance cycles are charted for $I_{comp}$ = 60uA, pulse set voltage = 1.5 V, pulse reset voltage = −1.3 V, pulse rise/fall time = 15 ns, and pulse top = 5 ns (**a**). The 50 cycle dc check after 344 billion cycles is shown in 12(**b**)

side. Only operating in the mode with a bias polarity having the positively biased reset-anode against the more stoichiometric part of the film (away from the OEL) results in well-behaved RRAM switching with well over 100 billion ac cycles demonstrated on device dimensions down to 50nmx50nm (see Fig. 3.12). Operating the device opposite this preferred polarity often results in the failure to reset toward higher resistance states (Fig. 3.11b) due to difficulty in reoxidation of the CF tip because of the high defect/vacancy density near the reset-anode in this "non-preferred" bipolar biasing case.

This oxygen vacancy asymmetry and preferred bipolar biasing is demonstrated further with a RRAM device using a decidedly asymmetric HfOx film formed by oxidizing the top of an exceedingly sub-stoichiometric HfOx film. The backside-SIMS profile of the oxygen concentration for the further oxidized HfOx film is shown in Fig. 3.13 indicating the film has high defect/vacancy concentration near the bottom TiN electrode and near stoichiometric $HfO_2$ near the top TiN electrode.

Note the defect profile for this stack in Figs. 3.13 and 3.14 (defects/vacancies near bottom electrode) is opposite to that of the engineered HfOx\OEL described above (with defects\vacancies near the OEL and top electrode). The ramifications of this opposite defect profile in the "top"-oxidized HfOx with respect to the HfOx\OEL ("top"-"reduced") stack, is that now the preferred bipolar operational biasing (i.e., having the positive reset-anode against the more stoichiometric side of the film), occurs with negative forming/set, and positive reset, voltages applied to top electrode. In addition, when performing the preferred operational biasing for this highly asymmetric "top"-oxidized HfOx, not only are over $10^8$ *ac-pulsed* endurance cycling and 100% *dc-iv* switching yield observed (Fig. 3.14), but operating with opposite this "preferred" bipolar biasing, with reset-anode against the defective/vacancy-rich side of the HfOx film, the resulting switching yield was zero%! Thus, RRAM

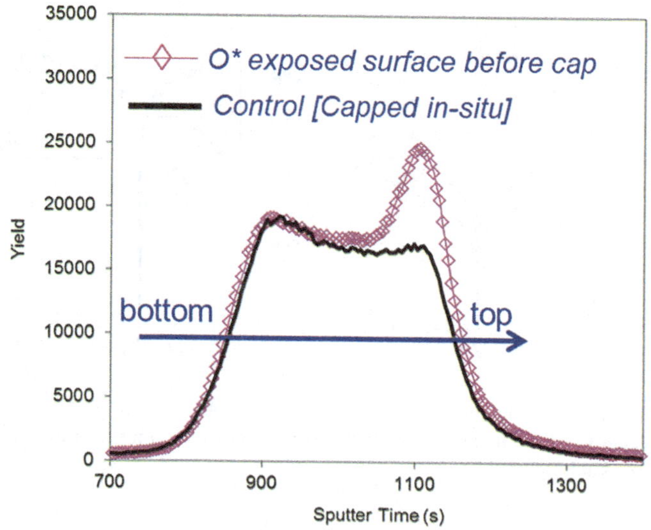


**Fig. 3.13** Backside SIMS showing the oxygen profiles for highly sub-stoichiometric $HfO_x$ with and without exposing the top surface to oxygen plasma before capping with Pt (Pt used only for SIMS analysis)

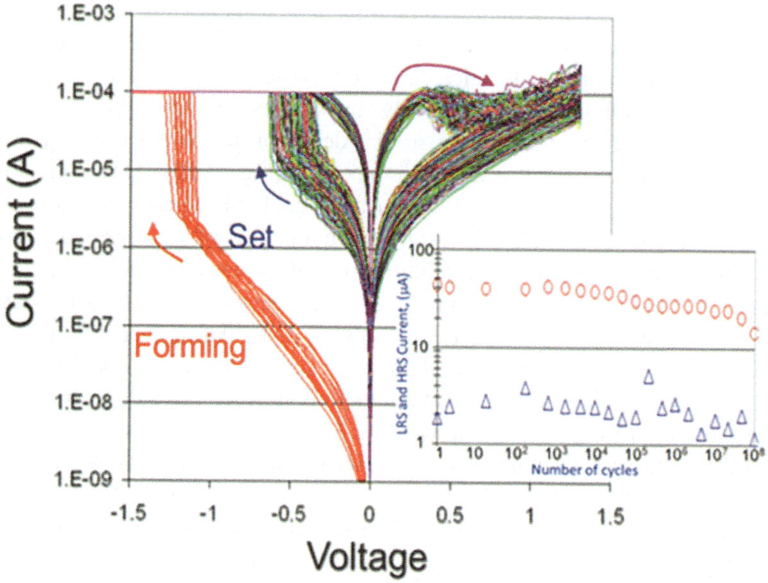


**Fig. 3.14** Over $10^8$ *ac* endurance cycling (inset) and 100% *dc* switching yield observed when performing the preferred operational biasing for a highly asymmetric oxidized $HfO_x$ film (with TiN electrodes and external compliance control). The dc switching shown is for 350 reset/set sweeps (for all 35 die tested x 10 cycles each)

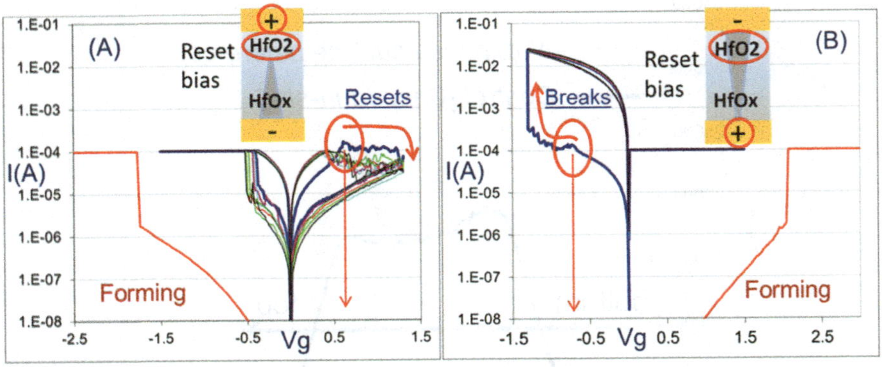


**Fig. 3.15** Typical dc forming and cycling with asymmetric $HfO_x$ for (**a**) preferred bipolar biasing (reset-anode against "good" more stoichiometric $HfO_2$) vs (**b**) non-preferred biasing (reset-anode against "bad" vacancy-rich $HfO_x$)

operation for this highly asymmetric sub-stoichiometric metal-oxide film dramatically illustrates the results of the "preferred" operational bipolar biasing relationship to the metal-oxide vacancy/defect profile and also how the oxygen vacancy profile supports the described reset mechanism of CF-tip oxidation. Having the positive reset-anode against the defective/vacancy-rich part of the HfOx stack allows for too many available conduction paths and insufficient oxidation of the CF for a HRS tunnel barrier to form. Furthermore, comparing *dc-iv* reset sweeps for the two bipolar biasing cases of reset-anode against "good" (more stoichiometric) or "bad" (more oxygen vacancies) part of the dielectric, it is observed that even for the "non-preferred" bias polarity with positively biased reset-anode against the "bad" vacancy-rich part of the dielectric, a similar deviation in current near reaching the forming compliance current value can be seen (Fig. 3.15). However, for this non-preferred biasing of the asymmetric vacancy stack, the current does not continue a downward trend to HRS (CF-tip oxidation) but begins to increase and the device "breaks" (with runaway current due higher rate of vacancy formation vs oxidation leading to a "breakdown," and with no compliance in reset, a resulting uncontrolled large CF is formed). Thus, the reset case with too many defects near the reset-anode suppresses the ability for CF-tip oxidation and denies a competition between filament oxidation and breakdown leading to only further breakdown (or CF size increase) and the resulting reset failure. Engineering the oxygen vacancy profile to support the described forming, reset, and set mechanisms is required for RRAM performance optimization.

Similar predisposition as the HfOx-based system for a preferred bipolar operation biasing and associated engineered oxygen vacancy asymmetry has been demonstrated for TaOx-based RRAM. Figure 3.16a shows *dc-iv* behavior for a TaOx-based stack consisting of a deposited thin layer of tantalum metal followed in situ by a deposited thicker and highly sub-stoichiometric TaOx layer, followed by a deposited thin stoichiometric $Ta_2O_5$ layer on top. The "preferred" bipolar operation of this stack is as expected with the positive reset-anode against the "good," or more

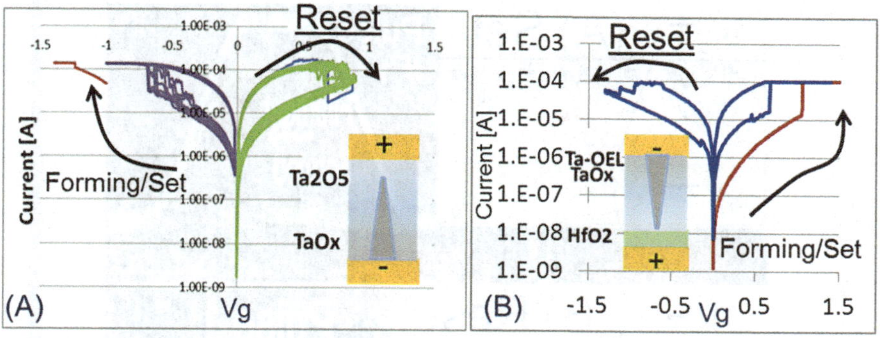


**Fig. 3.16** The dc forming/cycling for (**a**) asymmetric [bottom\Ta\$TaO_x$\$Ta_2O_5$\top] with "preferred" negative forming/set and positive reset applied to top electrode and (**b**) asymmetric [bottom\$HfO_2$\$TaO_x$\Ta-OEL\top] with positive forming/set and negative reset applied to top electrode

stoichiometric, "top" $Ta_2O_5$ side of the stack (so the forming was implemented with the negative cathode against the top electrode). Furthermore, a stack having near 1 nm stoichiometric $HfO_2$ deposited and then capped with a thick sub-stoichiometric TaOx layer, and then a thin tantalum metal-OEL layer (Fig. 3.16b) demonstrates similar bipolar switching as the RRAM stack with a single thicker $HfO_2$ with "top" deposited OEL of Figs. 3.10, 3.11, and 3.12 where preferred positive reset-anode is opposite the vacancy-rich OEL\top electrode (and against the more stoichiometric side of the stack). The similar reset for these devices with different metal-oxide RRAM material but similar vacancy profile stacks also supports the reset mechanism of oxidizing the CF tip near the reset-anode toward a thin (~1 nm) dielectric barrier [41]. The similar behaving films in Figs. 3.11a and 3.16b, which both have HfOx at the positive reset-anode, are both forming a similar ~1 nm $HfO_2$ tunnel barrier in reset with similar I-V characteristics [26, 45].

## 3.6 Real-Time Monitoring of RRAM Switching Highlights Operating Mechanism

These above-described relevant mechanisms for the bipolar metal-oxide RRAM operation can be highlighted by capturing the changes in the device conductivity in real time while applying "set or reset" type *ac* bias signals to the RRAM stack [6, 7, 26]. For instance, in general, a dielectric barrier limiting the current in a HRS should break similarly irrespective of polarity of the applied bias across the dielectric barrier; however, there is a clear difference to the HRS conductivity response when pulsed voltages of the same pulse width and height, but opposite polarities, are applied across the dielectric formed at the CF tip [26, 45]. For a "preferred" set-type bias (with a positive anode biasing near the defect\vacancy-rich side) applied to the device in the HRS, a clear abrupt change to a lower resistance state occurs

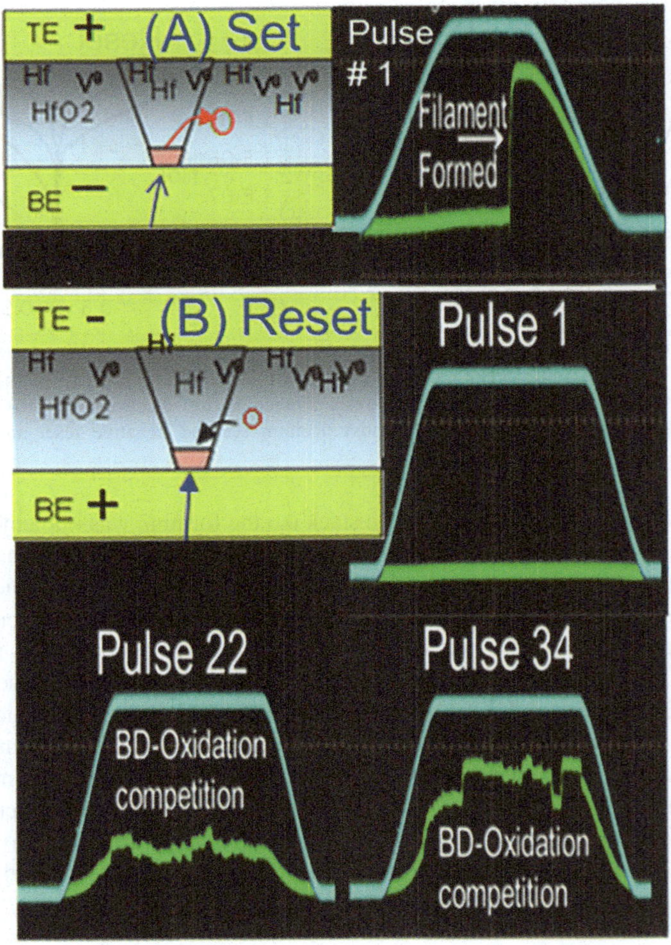


**Fig. 3.17** (**a**) Set pulse (100us and 0.8 V) applied to HRS state of the RRAM stack (similar RRAM stack as used for of Figs. 3.10, 3.11, and 3.12). Immediate and singular resistance change to LRS is observed indicating reformation of the CF for the set bias. (**b**) Reset pulse (100us and −0.8 V) applied to HRS state of the RRAM stack where many pulses are required for resistance change, and then the fluctuating resistance change is reflecting the competition between HRS breakdown and reoxidation

(Fig. 3.17a): the dielectric barrier abruptly breaks with the released negatively charged oxygen ions being pushed away from this dielectric barrier (that was formed at the CF tip in reset) and toward the positive anode by the electric field. Due to lack of "available" oxygen near the "breakdown" in the dielectric barrier and a higher oxygen vacancy concentration (from forming bias choice and the asymmetric oxygen vacancy engineering of the fabricated metal-oxide stack) near the positively biased anode for this "preferred" set-type biasing, the dielectric barrier breakdown occurs readily and immediately (and no dielectric barrier is formed at the opposite side where the positive voltage has been applied due to high vacancy concentration there). For the preferred reset-type bias (positive anode bias opposite the defect \vacancy-rich side and near the more stoichiometric side) applied to the device in the HRS, many continuous pulses are required before any resistance change occurs (Fig. 3.17b). In addition, when the resistance does change, the signal is very noisy with no immediate sharp current increase. According the above-discussed switching model, an apparent randomness of fluctuations of the HRS resistance observed under the continued "preferred reset-type" pulses results from a competition between the Hf-O bond breakage in the dielectric barrier and the near immediate reoxidation of the Hf atoms caused by a continued presence of the "available" oxygen ions in their vicinity: the applied field pushes these ions toward the positively biased electrode, thus preventing their fast out-diffusion from the filament region and providing ample available oxygen for filament tip oxidation in this area.

Figure 3.18 shows oscilloscope-captured switching data for a pulsed width of 20 ns for set, and 70 ns for reset, in fully integrated 1T1R $HfO_x$-OEL RRAM

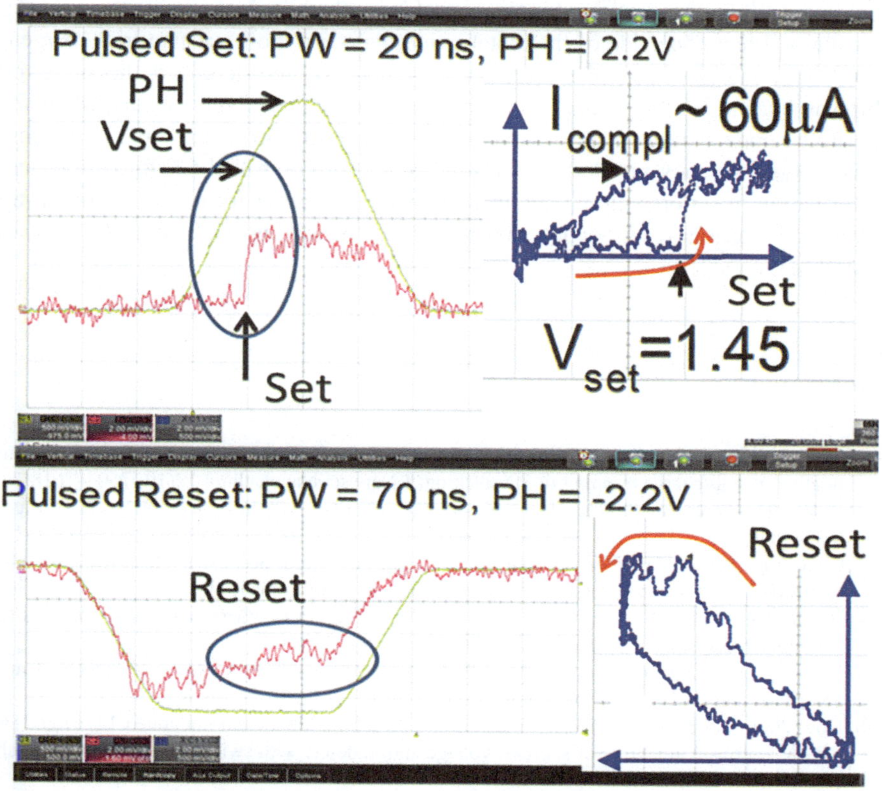


**Fig. 3.18** Pulsed switching in fully integrated 1T1R $HfO_x$\Ti-OEL RRAM for 100 nm × 100 nm cross-bar devices. Yellow, pulse voltage applied to the top electrode; red, voltage at the bottom electrode as read through the oscilloscope: inserts (blue), pulsed I-V traces corresponding to the respective pulsed set and reset operations

100 nm × 100 nm cross-bar devices (similar RRAM stack as used in Figs. 3.10, 3.11, and 3.12) [45]. The switching for both set and reset between HRS and LRS in the ns time frame is clearly observed. Additionally, for the "preferred" operational biasing applied, even at this ns time scale, the immediate and singular resistance change is observed for set operation (a dielectric breakdown), and the more noisy increase/decrease fluctuations in resistance are observed for reset operation (due to the reset bias additionally having an effective competition between oxidation of the CF tip to form a dielectric barrier and a breakdown of this developing dielectric barrier).

## 3.7 Conclusion

Optimizing metal-oxide RRAM switching requires an understanding of how engineering of asymmetric oxygen vacancy distribution and filament geometry assist the operation preferences of the bipolar switching schemes. In general, the vacancy profile will have one side of the metal-oxide dielectric more vacancy-rich/oxygen-poor with the opposite side more vacancy-poor/oxygen-rich, and the preferred bipolar biasing of this cell is such that reset more effectively occurs when the positive anode is next to the vacancy-poor/oxygen-rich side. The goal is creating this oxygen vacancy profile to be advantageous for controlled CF formation (and reformation in set operations), along with dielectric barrier formation (from CF-tip oxidation) during reset, under the conditions benefiting high-speed/low-power switching.

## References

1. Lu C-Y (2012) Future prospects of NAND flash memory technology-the evolution from floating gate to charge trapping to 3D stacking. J Nanosci Nanotechnol 12(10):7604–7618
2. Waser R, Aono M (2007) Nanoionics-based resistive switching memories. Nat Mater 6:833–840
3. Chen A (2011) Ionic memory technology. In: Kharton VV (ed) Solid state electrochemistry II; electrodes, interfaces and ceramic membranes. Wiley, [S.l.]. Cap. 1, pp 1–26
4. Bersuker G et al (2017) Toward reliable RRAM performance: macro- and micro-analysis of operation processes. J Comput Electron 16:1085–1094
5. Waser R et al (2009) Redox-based resistive switching memories – Nanoionic mechanisms, prospects, and challenges. Adv Mater 21:2632
6. Koveshnikov S et al (2012) Real-time study of switching kinetics in integrated 1T/ HfOx 1R RRAM: intrinsic tunability of set/reset voltage and trade-off with switching time. IEDM, Tech Dig
7. Koveshnikov S et al (2012) Development of NVM HfOx based 1T1R RRAM devices: needs and challenges for potential NAND replacement. (IGSTC) International GateStack Technology Symposium
8. Lee M-J et al (2011) A fast, high-endurance and scalable non-volatile memory device made from asymmetric $Ta_2O_{5-x}/TaO_{2-x}$ bilayer structures. Nat Mater 10:1–6

9. Lee HY et al (2010) Low-power and nanosecond switching in robust hafnium oxide resistive memory with a thin Ti cap. IEEE Electron Device Lett 31(1):44–46
10. Chen YS et al (2011) Challenges and opportunities for HfOX based resistive random access memory. IEDM Tech Dig:31.3.1–31.3.4. DOI:10.1109/IEDM.2011.6131649
11. Kim YB et al (2011) Bi-layered RRAM with unlimited endurance and extremely uniform switching. VLSI Technology, Symposium, pp 52–53
12. Yang JJ et al (2011) Metal/TiO2 interfaces for memristive switches. Appl Phys A 102:785–789
13. Butcher B et al (2011) High endurance performance of 1T1R HfOx based RRAM at low (20μA) operative current and elevated (150C) temperature. IIRW IEEE Trans Device Mater Rel (TDMR):146–150
14. Govoreanu B et al (2011) 10nmx10nm Hf/HfOx crossbar resistive RAM with excellent performance, reliability, and low-energy operation. IEDM, Tech Dig: 31.6.1–31.6.4. DOI:10.1109/IEDM.2011.6131652
15. Bersuker G, Jeon Y, Huff HR (2001) Degradation of thin oxides during electrical stress. Microelectron Reliab 41(12):1923–1931
16. Capron N, Broqvist P, Pasquarello A (2007) Migration of oxygen vacancy in HfO2 and across the HfO2/SiO2 interface: a first-principles investigation. Appl Phys Lett 91(19):192905
17. Foster S, Shluger AL, Nieminen RM (2002) Mechanism of interstitial oxygen diffusion in hafnia. Phys Rev Lett 89(22):225901
18. Mcpherson J et al (2003) Thermochemical description of dielectric breakdown in high dielectric constant materials. Appl Phys Lett 82(13):2121
19. Hubbard JK, Schlom DG (1996) Thermodynamic stability of binary oxides in contact with silicon. J Mater Res 11(11):2757–2776
20. Stemmer S (2003) Stability of gate dielectrics and metal gate electrodes. In: International SEMATECH gate stack engineering working group symposium, Austin
21. Stemmer S (2004) Thermodynamic considerations in the stability of binary oxides for alternative gate dielectrics in complementary metal–oxide–semiconductors. J Vac Sci Technol B: Microelectron Nanometer Struct 22(2):791–800
22. Cosnier V et al (2007) Understanding of the thermal stability of the hafnium oxide/TiN stack. Microelectron Eng 84(9–10):1886–1889
23. Guha S, Narayanan V (2009) High-κ/metal gate science and technology. Mater Res 39:181–202
24. Schaeffer JK et al (2003) Physical and electrical properties of metal gate electrodes on $HfO_2$ gate dielectrics. J Vac Sci Tech B: Microelectron Nanometer Struct 21(1):11–17
25. Bersuker G et al (2011) Metal oxide resistive memory switching mechanism based on conductive filament properties. J Appl Phys 110(12):124518
26. Bersuker G, Gilmer DC (2014) Metal oxide resistive random access memory (RRAM) technology, Chapter 9. In: Nishi Y (ed) Advances in non-volatile memory and storage technology. Woodhead Publishing, Cambridge, UK, pp 288–340
27. Kalantarian A et al (2012) Controlling uniformity of RRAM characteristics via the forming process. IRPS IEEE, Anahiem.
28. Gilmer DC et al (2011) Effects of RRAM stack configuration on forming voltage and current overshoot. IEEE-International Memory Workshop (IMW)
29. Kinoshita K et al (2008) Reduction in the reset current in a resistive random access memory consisting of NiOx brought about by reducing a parasitic capacitance. Appl Phys Letters 93:033506
30. Chen W et al (2015) Switching characteristics of W/Zr/HfO2/TiN ReRAM devices for multilevel cell non-volatile memory applications. Semicond Sci Technol 30(7):075002, https://doi.org/10.1088/0268-1242/30/7/075002
31. Fiorenza P, Polspoel W, Vandervorst W (2006) Conductive atomic force microscopy studies of thin $SiO_2$ layer degradation. Appl Phys Lett 88(22):222104
32. Porti M, Nafria M, Aymerich X (2004) Nanometer-scale analysis of current limited stresses impact on SiO2 gate oxide reliability using C-AFM. IEEE Trans Nanotechnol 3(1):55–60

33. Zhang L, Mitani Y (2006) Structural and electrical evolution of gate dielectric breakdown observed by conductive atomic force microscopy. Appl Phys Lett 88(3):032906
34. Porti M et al (2002) Electrical characterization of stressed and broken down SiO2 films at a nanometer scale using a conductive atomic force microscope. J Appl Phys 91(4):2071–2079
35. Yew KS et al (2009) Nanoscale characterization of $HfO_2$/SiOx gate stack degradation by Scanning Tunneling Microscopy. In: Proceedings of the international conference on Solid State Devices and Materials (SSDM), Japan
36. Mckenna KP, Shlugger AL (2009) Electronic properties of defects in polycrystalline dielectric materials. Microelectonics Eng 86:1751–1755
37. Fowler WB et al (1990) Hysteresis and Franck-Condon relaxation in insulator-semiconductor tunneling. Phys Rev B 41:8313
38. Vandeli L et al (2013) Microscopic modeling of electrical stress -induced breakdown in poly-crystalline hafnium oxide dielectrics. IEEE Trans Electron Devices 60(5):1754–1762
39. Gilmer DC et al (2002) Compatibility of polycrystalline gate deposition with $HfO_2$ and $Al_2O_3$/$HfO_2$ gate dielectrics. Appl Phys Lett 81:1288
40. Young-Fisher KG et al (2013) Leakage current – forming voltage relation and oxygen gettering in HfOx RRAM devices. IEEE – Electron Device Lett 34:750–752
41. Bersuker G et al (2010) Metal oxide RRAM switching mechanism based on conductive filament microscopic properties. IEDM Tech Dig:19.6.1–19.6.4
42. Bersuker, G. B. B. G. D. E. A. (2013) Connecting RRAM performance to the properties of the Hafnia-based dielectrics. ESSDERC. Bucharest: [s.n.].
43. B. Butcher, G. B. D. G. E. A. (2013) Connecting the physical and electrical properties of Hafnia-based RRAM. IEDM. [S.l.]: [s.n.]
44. Gilmer DC et al (2012) Engineering metal-oxide based RRAM for high volume manufacturing. Non-Volatile Memory Technology Symposium (NVMTS), Singapore
45. Gilmer DC et al (2012) Asymmetry, vacancy engineering and mechanism for bipolar RRAM. IEEE- International Memory Workshop (IMW), Milano
46. Stevens JE et al (2014) Reactive sputtering of substoichiometric $Ta_2Ox$ for resistive memory applications. J Vac Sci Technol A 32(2): 021501-1-6
47. Vallee C et al (2011) Plasma treatment of HfO2-based metal–insulator–metal resistive memories. J Vac Sci Technol A 29:041512
48. Hur JH et al Modeling for bipolar resistive memory switching in transitionmetal oxides. Phys Rev B 82:1–5 155321
49. Chien WC et al (2011) A multi-level 40nm WOx resistive memory with excellent reliability. IEDM Tech Dig. https://doi.org/10.1109/IEDM.2011.6131651
50. Lee HY et al (2008) Low power and high speed bipolar switching with a thin reactive Ti buffer layer in robust HfO2 based RRAM, IEDM. Tech Dig. https://doi.org/10.1109/IEDM.2008.4796677
51. Chen P-S et al (2009) Improved resistive switching of HfOx/TiN stack with a reactive metal layer and annealing. Solid State Devices Mater:444–445

# Chapter 4
# Organic Nanostructures by Molecular Layer Epitaxy: A Tutorial

**Tatjana N. Kopylova, Sergey Yu Nikonov, Evgeny N. Telminov, Ruslan M. Gadirov, Konstantin M. Degtyarenko, and Vladimir Burtman**

**Abstract** This chapter presents a short introduction to molecular layer epitaxy (MLE). MLE is a novel vapor-phase self-assembly approach to create molecular nanoelectronic materials, structures, and devices. The main principles of MLE technology and related topics of surface chemistry are introduced to demonstrate the possibilities of vapor-phase self-assembly, which can be used to achieve desired organic structures and chemical functionalities. We describe equipment, which can produce MLE structures, and monitoring methods used to describe growth of organic MLE structures. We demonstrate several devices, which were produced using MLE technology. These include field-effect transistors (MLE-FETs), organic light-emitting diode (MLE-OLED), MLE solar cells, and MLE laser media. The performance of MLE is compared with other technologies, which are currently used in molecular nanoelectronics, to demonstrate similarities and distinguish features of MLE.

## 4.1 Molecular Nanoelectronics

### *4.1.1 Molecular Nanoelectronics at Early Ages*

2017 marks the 20th anniversary of Mark Reed's article "Conductance of a molecular junction" on the possibility of electronic transport through a single molecule [1]. This was the first experimental step of molecular nanoelectronics, which was theoretically predicted by Aviram and Ratner in 1974 [2]. The initial expectations from molecular nanoelectronics were associated with ideas to combine advantages

T. N. Kopylova · S. Y. Nikonov · E. N. Telminov · R. M. Gadirov · K. M. Degtyarenko
Siberian Physics and Technical Institute, Tomsk State University, Tomsk, Russia
e-mail: kopylova@phys.tsu.ru

V. Burtman (✉)
University of Utah, Frederick Albert Sutton Building,115 S 1460 E #383, Salt Lake City 84112, Utah, USA
e-mail: vlad.burtman@utah.edu

 
S. M. Goodnick et al. (eds.), *Semiconductor Nanotechnology*, Nanostructure Science and Technology, https://doi.org/10.1007/978-3-319-91896-9_4

of cheap production of organic materials and their structural flexibility with the possibility to create single molecular semiconductor devices. In the early 2000s, experimental molecular nanoelectronics continued to produce vivid results. For example, the possibility of creating transistors with multiple stable states (based on the rotoxane molecule [3]) was demonstrated. Articles on spin transport in organic molecules discussed the possibility of creating an entirely new class of materials for molecular spintronics [4] and molecular magnetism. First experiments, which study organic molecules as biological objects, such as DNA, proteins, and neurons, can yield bioelectronic devices, which can transmit biological signals and control processes in living organisms [5].

Unfortunately, the initial expectations for rapid progress in molecular nanoelectronic devices have not come true. For example, Samsung indeed released few years ago a new generation of monitors, which are based on organic light-emitting diodes (OLEDs). However, the production costs of OLEDs monitors were similar to these models, which utilized inorganic semiconductors. Attempts to create efficient organic-based solar cells have not come true as well, due to their high costs and low operation performance. Currently only hybrid types of organic-inorganic devices, like Gratzel cells [6] and hybrid perovskites cells [7], still hold high expectations for device application. In addition, it was found that several key researchers in molecular nanoelectronics falsified the results of their work [8], and others misinterpreted them. For example, the interpretation of transport through individual molecules in Reed's article [1] was found to be erroneous [9].

This situation in molecular nanoelectronics at early ages was not accidental. Scientists working in molecular nanoelectronics concentrated their efforts on developing breakthrough device applications, rather than on fundamental studies of single molecular materials. As a result, molecular nanoelectronics still struggles to develop an adequate model of charge transport in molecular nanoelectronic devices based on a single molecular structure. For more than 25 years, scientists have been trying to adapt various models of solid-state physics, which were developed for inorganic semiconductors. These transport models relies on possibility to consider electrons in inorganic semiconductor devices as free by using the concept of effective mass of electron. The fundamental difference between organic and inorganic semiconductors is in absence of free charges in organic media. Due to this fact, there are no conduction and valence bands in organic semiconductors. Instead of conduction and valence bands, molecular electronics utilizes terms HOMO-LUMO. These acronyms stand for "highest occupied molecular orbital" and "lowest unoccupied molecular orbital," respectively. The applicability of free-electron theoretical concept is a problem in single molecular media. The fundamental structure-property relationship in molecular structures remains unclear, which impede development of new materials and devices in molecular nanoelectronics. Eventually, after exploration of easiest approaches and disappointments from lack of fast progress in molecular nanoelectronics, many scientists left this field. The remaining scientist began to redefine the broader scope of problems to solve in molecular nanoelectronics, began to utilize more complex scientific approaches to describe experimental objects, and, overall, began to employ higher scientific standards in their research.

### 4.1.2 Modern Times in Molecular Nanoelectronics

In 2013 the whole volume of Nature Materials [10] was dedicated to experimental studies in molecular nanoelectronics. These studies show a good example of new approach to molecular nanoelectronics, which is way more complex and sophisticated compared to experimental reports in early year's studies. Nowadays molecular nanoelectronics is shaping as classical science, in which the development of practical applications is combined with fundamental studies of the structure and properties in individual molecules.

The other distinguishing feature of modern time in molecular nanoelectronics is associated with the fact that this field received scientific and financial support from biology and medicine. In particular, new high-level studies on electronic transport in biological objects, as in DNA molecules, have been published [11]. Biological molecule, like DNA, has a similarity in structure to molecules used in molecular nanoelectronics. Understanding the mechanisms of charge transport in biology will make it possible to make significant progress in understanding similar mechanisms in molecular nanoelectronics.

In year 2016, Google and the British pharmaceutical company GlaxoSmithKline created a bioelectronic company for the use of miniature electrical implants in the treatment of chronic brain diseases, with the potential for further development of the interface between the brain and electronics [12]. In 2017, an American businessman Elon Musk entered into a similar project aimed at establishing a connection between the brain and the computer [13]. In the context of this development, we should note that only molecular nanoelectronics can provide a reliable interface between biological objects (neurons, DNA, proteins, etc.) and traditional electronic materials, including metal electrodes and semiconductor devices. In addition to biotechnologies, the new challenges for molecular nanoelectronics appear every year. These challenges include the development of catalytic nanomembrane technology [14], which can be utilized in new device concepts as an "artificial leaf" technology which is capable of converting exhaust gases into fuel using the energy of sunlight and can be entitled as artificial photosynthesis. The other example of contemporary challenges for molecular nanoelectronics is design and development of new field-effect transistors (FETs). Example of new FET system is shown in this book in chapter "The FinFET: A Tutorial" by Charles Dančak. Once single molecular transistor will become a robust and scalable technology, it has a good chance to become a new basis for electronic logics.

### 4.1.3 Quest for a New Fabrication Methods in Molecular Nanoelectronics

Among the numerous methods for creating materials with specified properties in molecular nanoelectronics, we should mention the methods of so-called "wet" and "vacuum methods." The wet methods include Langmuir films [15], self-assembling monolayers [16], and assembly of structures from solutions based on electrostatic interactions [17]. The vacuum methods include organic molecular beam epitaxy [18], other high-vacuum methods, and low-vacuum gas-phase deposition techniques, known as CVD technologies. Wet and vacuum methods have their advantages and disadvantages.

For example, "wet" methods are able to create stronger chemical bonds between organic layers, in differ to traditional methods of producing nanoelectronic structures from the "vapor" phase. However, the "wet" methods are very time-consuming and in their structure always contain a residual of solvents, which are able to affect the entire efficiency of the device. In addition, a number of interesting but insoluble organic substances cannot be used by "wet" chemistry.

Vacuum methods for creating stratified structures from the gas phase are capable of applying organic films quickly and in a clean atmosphere. They are fully compatible with modern semiconductor technologies. However, none of them is capable of creating strong covalent bonds between the layers, which limits the scope of the structures obtained. The main drawback of modern molecular nanoelectronic devices is instability. They are unstable for long periods of use, at elevated temperatures, with strong electric fields, unstable so that often even the spraying of electrodes presents a problem. The reason of these problems is the weak links inside the resulting films.

### *4.1.4 From Atomic Layer Epitaxy to Molecular Layer Epitaxy*

The main precondition for the device application of organic structures is the combination of epitaxial growth with covalent bonding in some similarity to the case of vapor deposition of inorganic semiconductors. The term *epitaxy* comes from combination of two Greek words: *epi* and *taxis,* which translates as "upon" and "arrangement." Therefore, the meaning of *epitaxy* is "ordering above," which simply means that the structure of the first layer defines the structure of the second layer. Many phenomena in chemistry, biology, crystal growth, geology, and material science are controlled by epitaxy; for example, the self-assembly process of fluids to form SAMs on metals. Some organic ultrahigh vacuum (UHV) [19] and organic molecular beam epitaxy (OMBE) methods [20] succeed in epitaxial growth, and some solution-derived SAMs can form interlayer covalent bonds. However, only inorganic atomic layer epitaxy (ALE) [21] can build covalently bonded epitaxial multilayers in a layer-by-layer fashion. The term *epitaxy* in ALE method means that the structure of upper layer is completely defined by bottom layer in ALE structure. While there are no strict terminology rules, the term *epitaxy* usually refers to the vapor-phase assembled structures, with strong (e.g., covalent) bonds between adjacent layers. When the structure is assembled using weak bonds (e.g., WdW bonds), the term *epitaxy* is used as a *pseudo-epitaxy* [16].

The ALE approach utilizes the difference between chemical and physical adsorption. When the first layer of atoms of a reactive molecular species reaches a solid surface, a chemical reaction (generally condensation) takes place, while subsequently deposited physisorbed layers tend to interact much less strongly. If the substrate surface is heated sufficiently, one can achieve a condition such that only the chemisorbed layer remains attached. Repeating this reaction cycle with different compounds leads to a controlled layer-by-layer growth. The ALE method was successfully applied to the epitaxial growth of "conventional" semiconducting inorganic materials, such as II–VI and III–V compounds and silicon and silicon-related compounds (see molecular beam epitaxy, atomic layer epitaxy).

A new method, molecular layer epitaxy (MLE) [22], has been developed for the deposition of organic heterostructures. In contrast to most ordered organic SAM thin films that are assembled in solution, the MLE approach employs a low-pressure vapor-phase epitaxial growth via covalent bonds between organic interfacing layers. Term epitaxy in MLE method designates organic structures based on strong bonds, which bonding energy above 1.8 eV.

## 4.2 Methodology of Molecular Layer Epitaxy

### *4.2.1 MLE Principles and Hardware*

The MLE method is based on four principles: a template layer, self-restricted vapor-phase reactions, covalent ("c-axis") interlayer bonding, and π-stacking in the x-y plane (the plane of the substrate). The MLE method is based on the use of modern equipment for rapid and clean application of organic layers (Figs. 4.1 and 4.2), but unlike other "gas" methods, MLE allows for chemical self-assembly of layers with the formation of strong covalent bonds between them. The technology is based on the successive growth of a layer on the surface of a substrate from individual molecules of vapors of a heated organic substance. Specially designed for MLE vacuum system was developed and assembled on the basis of the most modern thin film technology. This unit heats the precursor materials to high temperatures

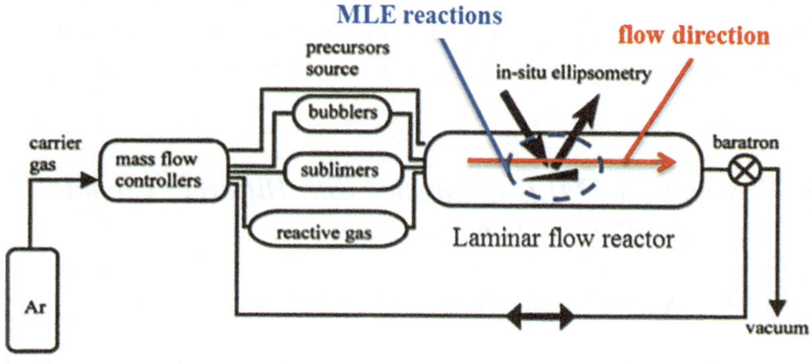


**Fig. 4.1** Scheme of laminar flow MLE reactor setup

**Fig. 4.2** Photograph of 2nd generation of molecular layer epitaxy setup

(500 °C) in a vacuum ($10^{-5}$ Torr). The evaporated molecules are trapped in a special way and placed on the substrate layer by layer, forming the thinnest film (*c.a.* 3 nm).

The second generation of MLE setup (Fig. 4.2) inherent from the first generation of MLE setup [22] the similar shape of CVD chamber and position of its functial elements (sublimators, bublers, pumps etc). It featured an addition of few high-temperature sublimes and higher reactor temperature and pressure. These additions enable usage of high-temperature precursor materials and using vapor-phase self-assembly reactions at higher temperatures. The ultimate goal of 2nd generation MLE setup was to gain a realistic possibility to include inorganic precursor materials in MLE chemical routine to produce mixed organic-inorganic superlattices within the same reactive chamber. The other additions to 2nd generation of MLE setup includes better temperature, pressure and evaporation control, and an easy-to-set automatic operation mode to illuminate necessity in direct manual operation of MLE process.

These artificial submicron organic structures are characterized by a high degree of ordering and strong covalent bonds between the layers, as shown in Fig. 4.3.

### 4.2.2 *Examples of MLE Chemistry with Different Chemical Functionality on Top*

**MLE with Top Amino-Group Chemical Functionality**

The surface chemistry of the MLE method, as shown in Fig. 4.4, allows the growth of naphthalene-diimide-based organic structures.

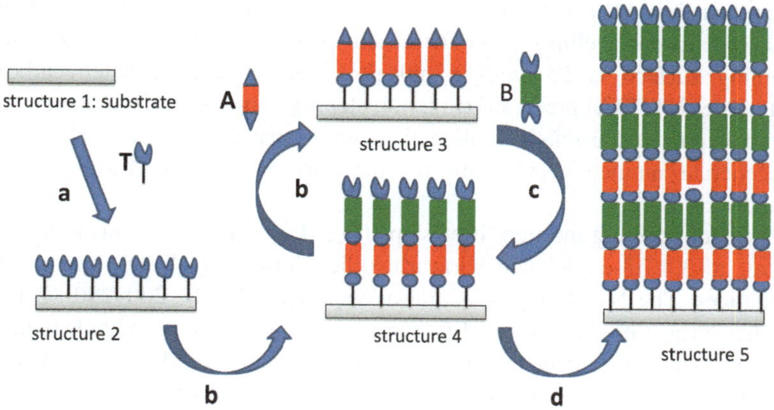


**Fig. 4.3** Molecular layer epitaxy (MLE) method: organization of structures, when the structure of the upper layer is controlled by the structure and chemical composition of the previous layer. The figure shows the scheme of epitaxy. First, the "anchor layer" (arrow "a") is placed on substrate 1 (the arrow "a"); structure 2 is formed. Next, a layer of molecules A (arrow "b") is applied on 2; a structure 3 with two assembled layers is formed. Next, a layer of molecules B (arrow "c") is applied; a structure 4 is formed. Thus, it is possible to assemble a structure of necessary thickness with alternating layers (arrow "d")—a superstructure 5 is formed. It is essential that layer A is adapted to accept layer B, and vice versa, it is impossible to collect more than one layer only A or only B

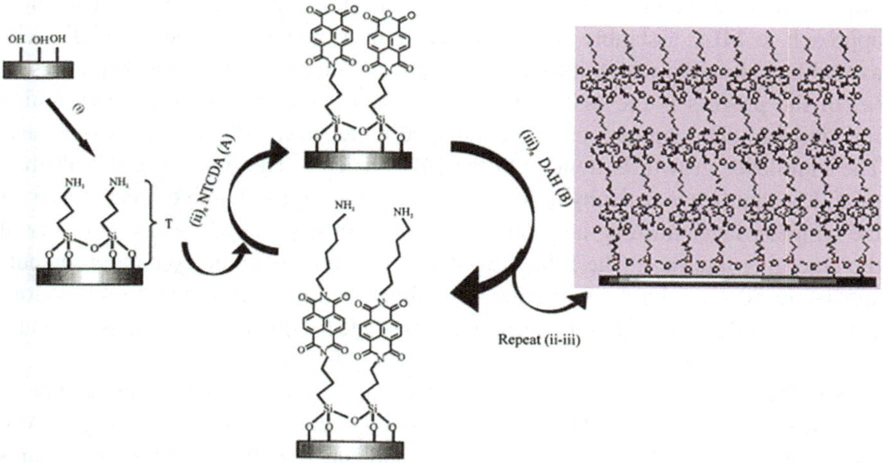


**Fig. 4.4** MLE surface chemistry with amino group on top of MLE structure

A template layer is first deposited on oxide surfaces, such as Si/SiO2 (Fig. 4.4, step i), exposing propylamine functionality toward the interface, and in turn this layer dictates epitaxial growth. Then discrete pulses of reactants —liquids or solids that undergo self-limiting reactions—are carried to the surface by an inert carrier gas in a laminar flow MLE reactor setup (Reynolds number ~250) (Fig. 4.2) equipped with a spectroscopic ellipsometer for in situ process monitoring. An alkylamine-containing surface is hit with a pulse of 3,4,7,8-naphthalenetetracarboxylic-dianhydride (NTCDA) precursor (Fig. 4.4, step ii), forming imide linkages. Then a pulse of a vaporized aromatic or aliphatic diamine spacer (Fig. 4.4, step iii) is added to regenerate amine functionality on the surface that can again react with dianhydride.

Repeatedly cycling through these steps (Fig. 4.4, steps ii,iii) leads to the formation of a 3,4,7,8-naphthalenetetracarboxylic diimide (NTCDI)-based organic superlattices. The NTCDA precursor was vaporized at T = 200–250 °C to react with the surface-bound amine; 1,6-diaminohexane and 4,4«-methylenedianeline precursors were vaporized at T = 35–40 °C and 90–110 °C, respectively. Imidization in both steps is carried out at a substrate temperature of 280–290 °C for 15–25 min at a total pressure of 0.05 torr. Following the assembly of each monolayer, the reaction zone is cleaned by resublimation of unreacted precursors under $N_2$ flow for 5 minutes. The reactor walls are independently heated to 300 °C.

## MLE with Top Thiol Group Chemical Functionality

Fabrication of different molecular electronic devices usually requires different chemical functionality. In particular, thiol functionality is often required for formation of chemical bonds with upper metal electrodes (Au, Ag, Al, etc.). Chemical approach to MLE assembly of naphthalenetetracarboxylic diimide (NTCDI) self-assembling structure with thiol group on top is highlighted in Fig. 4.5. The first step *(a)* of the process is the assembly of a template layer containing primary amine functionality. This process can be conducted either from solution or vapor phase. The multilayer growth is done with the MLE reactor. The first pulse of substrate (Fig. 4.5, step *(b)*) is a dianhydride reaction with the exposed amine functionality of the template layer forming an imide bond and exposing anhydride functionality at the emerging interface. The following step *(c)* anchors diamine containing precursors (spacers) again by the formation of an imide bond and exposing thiol functionality on top. Molecular 2D surface aggregates, stage *(d)*, grow orthogonally to planar view of NTCDI molecule, shown between stages *(b)* and *(d)*.

NTCDI-based structures were accomplished as follows. Thick, impermeable to light, wafers of n-type (500 Ω-cm, Virginia Semiconductors) Si(100) having 20 Å of $SiO_2$ coating and glass slides were cleaned and functionalized with amino groups (Fig. 4.5, *step a*) as described in ref. 17. NTCDA (1,4,5,8-naphthalene tetracarboxylic anhydride) evaporated at 110 °C and reacted with the $NH_2$-functionalized surface for 45 min at $10^{-5}$ Torr (Fig. 4.5, *step b*). The product of assembly lacks the amino group preventing formation of the second layer. The substrate was kept on a heated sample holder (180 °C) preventing physadsorption of the precursor. Vacuum deposition modified over 60% of the surface amino groups [17]. A layer of 4-aminophenylthiol was added within 20 min (Fig. 4.5, *step c*). This reaction tops the surface with SH-groups reactive toward metals [23].

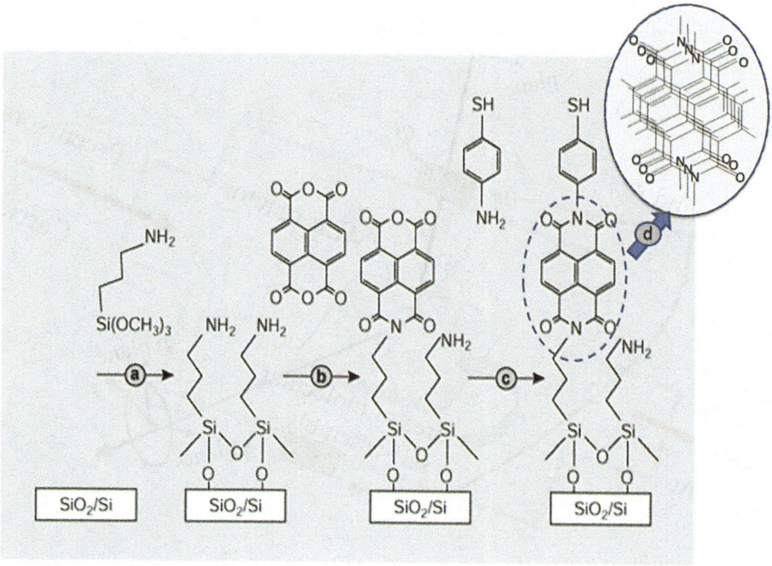


**Fig. 4.5** Surface chemistry, which yields surface π-aggregation system in naphthalenetetracarboxylic diimide (NTCDI) system by molecular layer epitaxy (MLE) method with a thiol group on top. The NTCDI aggregates may be visualized as structures that are orthogonal (*out-of-plane*) to picture plane, as shown in right upper corner

The control experiments to verify the quality of vapor-phase self-assembly include spectroscopic in situ and ex situ ellipsometry, measuring progressive contact angle, XPS, and recording of absorption spectra after deposition of each layer [24].

### 4.2.3 *Monitoring Tools to Trace the Growth of Organic Superlattices*

#### Monitoring of Growth of Organic Superlattices by Spectroscopic Ellipsometry

Experimental setting for in situ spectroscopic ellipsometry monitoring the MLE process is shown in Fig. 4.6. This figure highlights position of Woollam M44 spectroscopic ellipsometer, including position of optical windows, substrate, and reactive zone, in respect to axis of MLE reactor, precursor and carrier gas flow, vacuum, and unreacted products. The ϕ angle is a spatial angle between incoming linearly polarized light and orthogonal to reflection point on substrate in the plane of incidence. Linearly polarized light become circularly polarized after reflecting from MLE structure, which is grown on the substrate in real time (Fig. 4.6). In ellipsometry light, which vector of electrical field oscillated within the plane of incidence is called *p*-polarized light, and light, which vector of electrical field

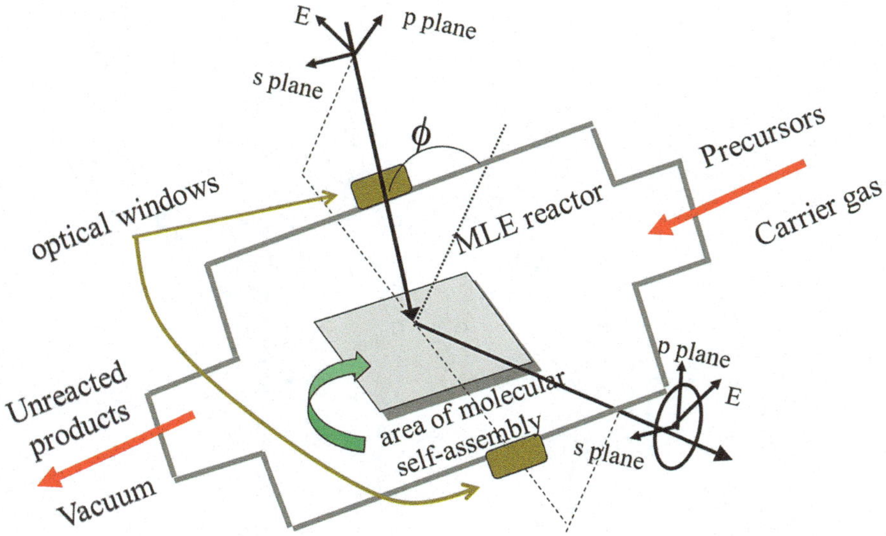


**Fig. 4.6** Experimental setting for in situ spectroscopic ellipsometry monitoring the MLE process

oscillated perpendicular to the plane of incidence is called *s*-polarized light. The basic principles of ellipsometry rely on the fact that *s*- and *p*-polarized light reflect differently: $R_p \neq R_s$. The ellipsometry measures the complex reflectivity ratio, $R_p/R_s = \tan(\psi)e^{i\Delta}$, and typically reports results in terms of ψ and Δ parameters. The tan (ψ) is a magnitude of the $R_p/R_s$ ratio and Δ is a phase difference between *p*- and *s*-reflected light. We study the evolution of ellipsometry parameter ψ, which changes in real time and reflects the change of light polarization upon different stages of self-assembly reaction and growth of organic superlattice.

Figure 4.7 shows an example of MLE process monitoring for continuous growth of naphthalenetetracarboxylic diimide-hexamethylene $(NTCDI\text{-}HM)_n$ structure, $n = 4$, assembling from NTCDA and DAH precursors in single wavelength representation. This figure shows the change in the surface polarization parameter ψ, detected by in situ ellipsometry versus deposition time.

The temporal evolution of the monolayers by in situ ellipsometry provides several interesting observations. First, the ellipsometry parameters ψ and Δ are very sensitive to the surface prefunctionalization. Even the raw data exhibit the periodical growth and decay of surface polarizability at different steps of the ML-assembling process with periods of about 15 minutes for the NTCDA pulse and 18 minutes for the DAH pulse. Figure 4.7 also provides information on the kinetic behavior of the self-assembly process since ψ and Δ can be converted to surface chromophore or spacer-relative coverage. The rate coefficients k for every pulse are $5.6 \times 10^{-5}\ s^{-1}$ for first deposition of NTCDA on a template layer, decreasing to $1.5 \times 10^{-5}\ s^{-1}$ for NTCDA on a hexamethylene layer and to $7.2 \times 10^{-6}\ s^{-1}$ for DAH on a naphthalene layer with an equilibrium surface absorption energy 15.8 kcal for the NTCDA layer and 29.2 kcal for the DAH layer.

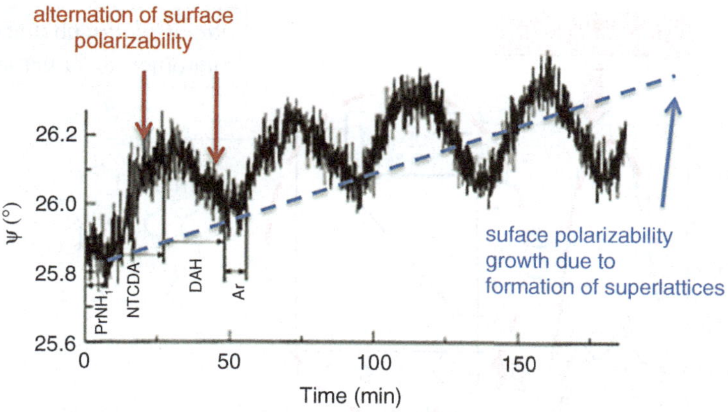


**Fig. 4.7** In situ monitoring of MLE monolayer-by-monolayer growth

The in situ (Fig. 4.7) and ex situ [19] ellipsometry measurement is a very useful tool in monitoring of MLE-derived structures, especially for cases in which different building blocks of organic superlattice have a different molecular polarizability, as in case of NTCDI-HM system.

The ex situ spectroscopic ellipsometry can be useful for structural analysis of MLE superlattices [19]. However, ex situ spectroscopic ellipsometry is less illustrative and requires knowledge of optical constants ($\varepsilon$ and $n$) for each consistent layer to build an adequate model for MLE structure.

## Monitoring of Growth of Organic Superlattices by Optical Spectroscopy

Monitoring of growth of organic superlattices by optical spectroscopy is one of the simplest and very sensitive tools to ensure a consistent growth of MLE-derived organic superlattices. The precondition for using this monitoring method consists in usage of sensitive spectrophotometer with spectral resolution of optical density (OD) about 0.001. This resolution has to resolve the growth of one monolayer. The other technological complication for this (ex situ) method is the necessity to interrupt MLE process to record Vis-UV spectra and then to continue MLE process. This approach might be problematic for a range of materials and technologies, including, for example, air-sensitive materials. In situ reflection optical spectroscopy monitoring should resolve this complication. However the alignment of in situ optical instrumentation might be complicated, especially taking into account the necessity of high OD sensitivity.

The structure of organic superlattices, which is traced by ex situ *UV-Vis* optical spectroscopy, is shown in Fig. 4.8. This figure provides an easy access to understand the main spectral features, which should be traced to ensure the quality self-assembly in vapor phase. Figure 4.8 shows a monolithic growth of the ten bilayer superlattice containing aliphatic spacer in the absorption at 362 nm. We designated this band as a

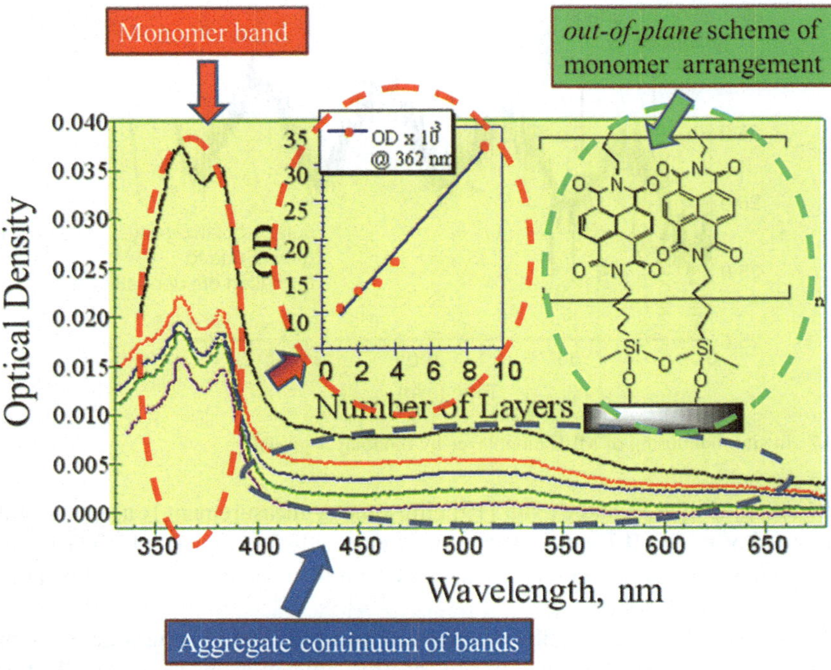


**Fig. 4.8** The structure of organic superlattices, which is traced by UV-Vis optical spectroscopy: absorption spectra of MLE-derived $(NTCDHI)_n$-based organic superlattice for n = 1,2,3,4,9. Different spectral features are specified in form of colored dashed lines, arrows, and text boxes

"monomer band" (red text box, arrows, and dashed lines). The middle panel in Fig. 4.8 shows the monotonic growth (blue line in the middle insert) of optical density at 362 nm of this "monomer band." The *out-of-plane* structure of growing NTCDI-organic superlattice is shown in the right upper insert (green dash line and green text box). Note that for graphic simplicity we do not show here the *in-plane* order in NTCDI-organic superlattice. The in-plane packing inside one NTCDI layer in NTCDI superlattice is shown in upper right part of Fig. 4.5.

The intermolecular charge transfer (CT) band is observed around 450–700 nm exhibiting the high in-plane packing density. We designated this band as an "aggregate continuum of bands" (blue text box, arrows, and dashed line). This band is originated from delocalization of π-electrons in aromatic rings during aggregation of NTCDI moieties into clusters, in which these electrons become pseudo-conductive in narrow continuum of bands (COB) [25].

To verify the quality growth of organic superlattices, the main attention has to be paid to tracing of growth of OD of monomer band (362 nm, Fig. 4.8). These spectra are recorded after each step of self-assembly reaction and measuring the absorption spectra. The growth of OD should be of order of ~ 0.001 for addicting of next layer, this growth should be consistent and such growth should be linear in first approximation. Any inconsistency in monitoring spectra in form of much lower or much higher increment of OD for any grown layer, or change in shape of monomer band, or sharp shift of monomer band (362 nm) to blue or red region is a clear indication of problems in growth of organic superlattices. Most common problem in MLE process is interruption in growth of OD, which indicates problem in MLE growth. Despite difficulties, which are noted in the beginning of this section, the monitoring of growth of organic superlattices by optical spectroscopy is still very efficient for adjustment stage of MLE technology for any new material system.

The distinguished feature of spectroscopic ellipsometry and UV-Vis spectroscopy consists in their capability to trace not only the fact of molecular self-assembly in vapor phase but also to evaluate the molecular density in each consistent layer. This is particularly useful to avoid the problem of pyramidal growth, which is typical for solution molecular self-assembly.

### Other Monitoring Methods Used to Describe the Grow of Organic Superlattices

The monitoring of nonlinear response from growth organic structures can be used to trace in real time as well in certain cases. For example, the measurement of second harmonic generation (SHG) response from organic superlattices, which contain the molecules with SHG response, can be used to trace their growth [26]. The list of simple and sensitive tools to trace the growth of organic superlattices includes measurement of progressive contact angle after each step of self-assembly reaction. FTIR spectroscopy is another useful tool to trace chemistry of substituted amine (substitution of oxygen in NTCDA ring to nitrogen) during MLE growth [22]. These monitoring tools today cannot be used directly to evaluate the package density within each consistent layer and, therefore, can be employed as quality tools.

We impose in this section that increment in spectroscopic and polarization response from growing organic superlattice can be considered additive in first approximation. This includes the monotonic growth of polarization and optical densities in Fig. 4.7 (blue-dashed trend line) and Fig. 4.8 (blue trend line in the middle insert of optical density at 362 nm). This assumption is well enough for a general monitoring purpose. However, the close look on dependence increment in spectroscopic and polarization response from growing organic superlattice (e.g., growth of OD with number of monolayers, $n$, see Fig. 4.8) reveals nonlinearity at $n > 10$. This nonlinearity is attributed to size-dependent effects in organic superlattices. The measurement of size-dependent effects in organic structures using absorption spectroscopy along with electro-optical properties of MLE structures is discussed in the following section.

### 4.2.4 Size-Dependent Effects in MLE Structures

Accounting the size-dependent effects in MLE structures might be critical in some situations, for example, when a good *in-plane* density of MLE films becomes critical and the number of contained monolayers $n$ exceeds ten ($n > 10$). Example of devices, which may contain such structures, includes planar photovoltaic cells, with capability to harvest light excitation from macroscopic distances, or MLE-FET devices. In these cases the size-dependent effect should be accounted to ensure an adequate accuracy of structural evaluation.

However, this remarkable feature of superlattices is a topic of independent scientific interest. The size-dependent characteristic in NTCDI-based superlattices can be studied by their optical absorption signature. The blueshift in the absorption peak corresponding to an increase in transition energy ΔE, predicted as the thickness of the multilayer decreases, was observed for a range of organic and inorganic superlattices. The same trend is observable in the lowest-energy absorption peak of the MLE-derived NTCDI superlattices, depicted in Fig. 4.9a. An energy decrease of about 12 meV is measured for the multilayered structures containing the fewest layers of NTCDI with aliphatic HM spacers. Considering the organic multiple quantum well (OMQW) energy levels to consist of NTCDI (a) semiconducting narrow potential wells placed in between very high-potential barriers (b) (aliphatic spacers), both electron and hole can be approximated as confined in the NTCDI layer. Then a model of a confined Wannier exciton in coupled quantum wells can be considered. The exciton binding energy ($E1s = \langle\psi|H|\psi\rangle$) was determined to fit the blueshifted absorption peak, shown as a dashed line in Fig. 4.9a.

Variety of other spectroscopic tools can be employed to study the growth of MLE structures. For example FTIR spectroscopy is capable to trace the chemical reaction, in which surface amine react with NTCDA anhydride to produce the imide interlayer bonding [27].

## 4.3 MLE-Derived Molecular Nanoelectronic Devices

### 4.3.1 MLE Field-Effect Transistors (MLE-FETs)

The MLE-derived devices' operation stability is robust and able to operate at a DC field of $3 \times 10$ Vcm$^{-1}$. This stability, unusual for organic devices, can be explained by the energy of imide bonds that is about 1 eV in contrast to that of van der Waals bonds (0.01 eV) achieved by most solution and ultrahigh vacuum (UHV) deposition methods. The possibility to achieve controlled epitaxial ML growth and ohmic contacts on an inorganic-organic level enables us to fabricate stable and highly ordered molecular electronic devices as OFETs and organic single-electron

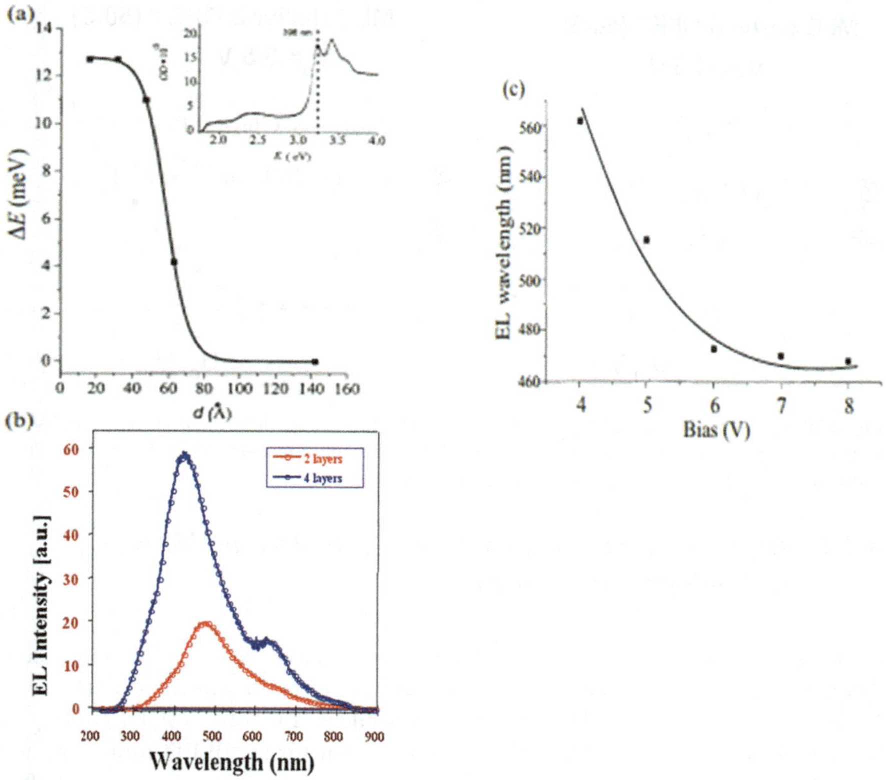


**Fig. 4.9** Scaling laws for MLE-derived organic superlattices: (**a**) Energy shift of the NTCDI-HM superlattice absorption peak, marked by the arrow in the insert graph, as a function of thickness. The symbols correspond to experimental data, and the solid curve is the fit to model described in the text. Insert: Absorption spectrum of (NTCDI-HM)$_2$ multilayer sample. (**b**) Thickness-dependent EL in (NTCDI-DPM)$_n$ for n = 2 and 4 in MLE-based OLED at VDC = 8 V. (**c**) DC voltage-dependent EL for (NTCDIDPM)$_4$-based OLED

transistors. 1,4,5,8-Naphthalene-tetracarboxylic diphenylimide (NTCDI)-derived compounds were already tried in organic devices, including thin and thick film OFETs. The highest mobilities of solution-processed n-type semiconductors were achieved from quaterthiophene (0.2 cm$^2$ V$^{-1}$ s$^{-1}$) measured in a vacuum and from naphthalene diimide (NTCDI) with a long-chain fluorinated alkyl group (0.01 cm$^2$ V$^{-1}$ s$^{-1}$) solution-processed in air. The highest n-channel mobilities of these devices, which were based on NTCDI derivatives, were calculated to be about $1.2 \times 10^{-3}$ cm$^2$ V$^{-1}$ s$^{-1}$. A room-temperature electron mobility as high as 90 cm$^2$ V$^{-1}$ s$^{-1}$ was reported using MLE. Figure 4.10 shows the preliminary results of the characteristics of nano-organic field-effect transistors.

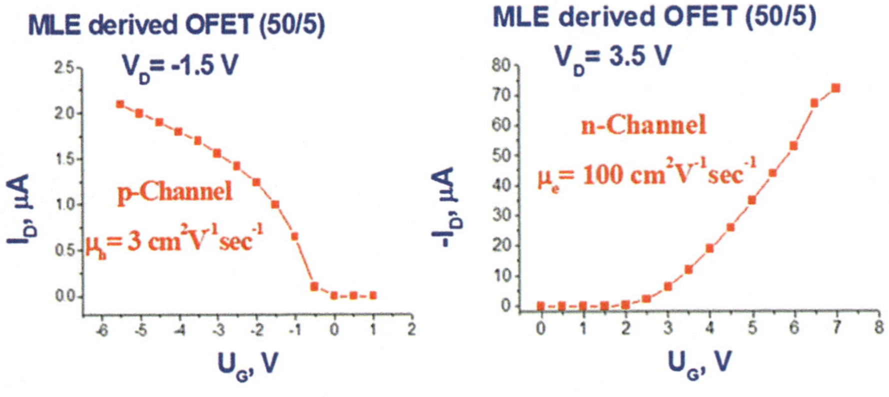


**Fig. 4.10** Mobility charges in bipolar field-effect transistors (nano-OFETs), produced by the MLE technology. The mobility of electrons in the MLE structures reaches 40–200 $cm^2/Vs$, which exceeds the best known values for organic transistors

## *4.3.2 Organic Light-Emitting Diodes produced by MLE technology (MLE-OLEDs)*

MLE-derived OLEDs, as small as 4–6nm thick active layers, were fabricated. Another finite size effect was observed in electroluminescence (EL) of MLE-derived structures. The thickness-dependent EL of $(NTCDI\text{-}DPM)_n$ is depicted in Figure 4.9(b). The EL spectra is shown for an OLED formed by the assembly of 2 and 4 bi-layered NTCDI structures with diphenylmethane (DPM) spacer in a glass/ITO/organic multiply quantum well (OMQW)/Al configuration at a bias of 8V, where ITO stands for indium tin-oxide electrode. The 480nm EL maxima for $(NTCDI\text{-}DPM)_n$ was shifted almost 50 nm with doubling of the thickness, and it was clearly observed that two different centers emit under these conditions. The blue EL is attributed to radiative recombination on a single NTCDI molecule and the red EL is assigned to radiative recombination within aggregated NTCDI molecules. The coexistence of two emitting centers within the same layer is an indication that two different excitations exist: (i) intra-molecular excitons and (ii) inter-molecular excitons. The same trend of a new PL peak appearing due to epitaxial ordering (intermolecular excitons) was observed in a single crystal of naphthalene tetracarboxylic di-*n*-hexylimide (NTC- DHI) grown by the Bridgman method. The EL can be shifted to the blue by ramping the applied potential in the same $(NTCDI\text{-}DPM)_n$, OMQW based devices (Figure 4.9(c)). At low voltage, the red emission of the stacked NTCDI molecules (intermolecular excitons) dominates the EL spectrum, since these centers have a lower energy gap for excitation and charge recombination. Increasing the applied potential leads to charge recombination at the higher energy gap, intramolecular excitons, and to blue-shifted EL. In summary, finite size effects govern blue-shifting of absorption peaks and DC-field EL tunability.

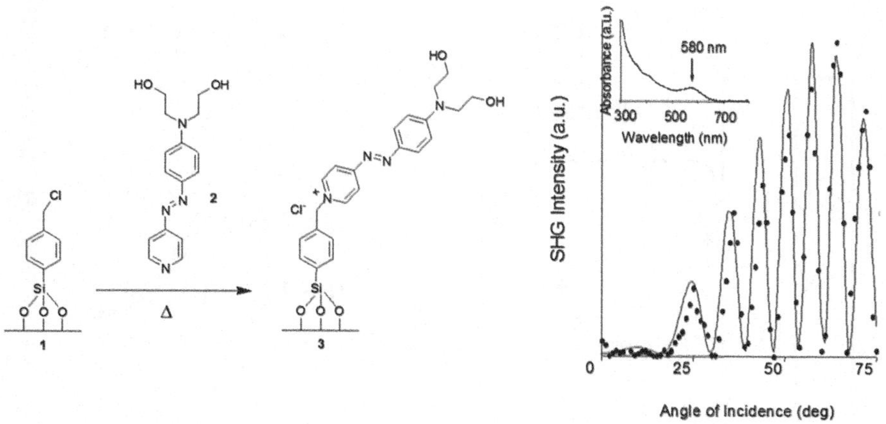


**Fig. 4.11** Example of laser action in MLE structures. Left is a molecular self-assembly in the vapor-gas phase of a precursor (amino-phenyl-azo-pyridinium) on a modified glass substrate. On the right is the generation of the 2nd harmonic upon excitation by light of a wavelength of 1064 nm. The spectrum of the assembled structure is shown on the right side of the inset

### 4.3.3 Laser Media Produced by MLE Method

An example of laser action in MLE structures is shown in Fig. 4.11.

This structure cannot be obtained by liquid-liquid methods due to the insolubility of the precursor (amino-phenyl-azo-pyridinium) and vacuum methodologies that are not capable of creating a chemical bond between the fungicide-bonded silicone substrate and the active molecule. The main characteristic of MLE method is the ability to create stable structures that, from the point of view of their stability, behave in the same way as their inorganic counterparts (e.g., atomic layer epitaxy method).

### 4.3.4 MLE Photovoltaic Cells

The photovoltaic properties of MLE-derived NTCDI structure are exhibited in Fig. 4.12 [28]. The NTCDI structure, which featured in-plane molecular stack system, was assembled on boron-doped silicon wafer. Device configurator is shown at Fig. 4.12a. This measurement enables to evaluate the velocity of redox electron transfer in direction of molecular NTCDI stack (reaction 2 in Fig. 4.11a) and demonstrates the possibility to harvest photo-excited charges at cm-scale distance.

The photovoltage transients consist of voltage rise and decay (Fig. 4.12). The decay is exponential with $\tau \sim 40$ μs at the given external load. This component represents the discharge of the cell because of both the backflow and the functional current through the load. The voltage rise originates from accumulation of charges at the anode, and its kinetics depends on the distance between the illuminated spot and silver (Fig. 4.12). To avoid a model bias, we described the kinetics of the voltage rise

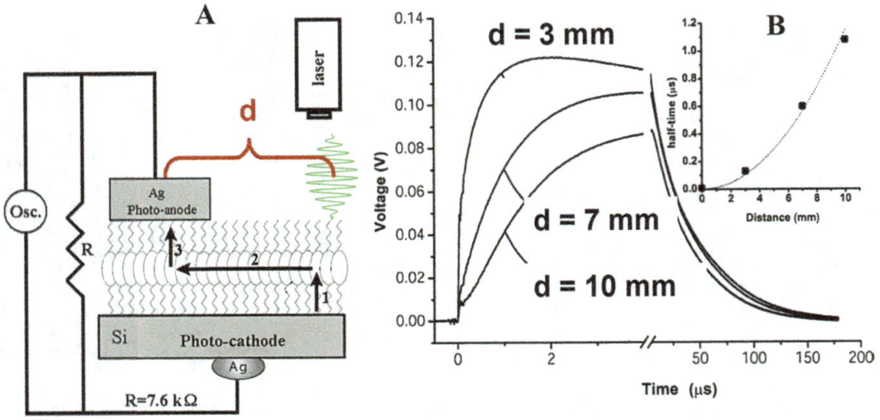


**Fig. 4.12** Scanning time-resolved photovoltage probing. (**a**) Measurement setup. Distance *d* is between illumination point and harvesting upper Ag electrode. The transients were induced by pulses centered at different distances apart from the silver electrode. (**b**) Kinetic of photoresponse at different *d*. The inset shows the dependency of the $T_{1/2}$ of the voltage rise on the distance between the bright spot and the silver photoanode. The solid line approximates the points by a parabola $T_{1/2} = l^2/D$ with the parameter D~$10^6$ cm$^2$/s

by its halftime $T_{1/2}$ (Fig. 4.12, inset), which reflects the travel time of the cation radicals. Simple estimate shows that corresponding speed of charge exchange exceeds $10^4$ m/s. The mutual arrangement of the cell and the incident light; the external quantum efficiency (EQE) of 0.4–0.7) [23], which cannot be accounted for by the absorption efficiency of the film; the dissimilarity of the EQE; and the absorption efficiency of the film and resemblance of EQE to the spectrum of silicon suggest [23] that the light is productively absorbed by the semiconductor (Fig. 4.12). This contrasts with light harvesting in the dye-sensitized photovoltaic cells [29, 30] or the hybrid nanorod-polymer cells [31], which require special junctions with enormous contact areas between the materials to enhance their absorption efficiency. The geometry of the cell also suggests that light harvesting incorporates a longitudinal spatial energy transfer. In contrast to natural photosynthesis [32], the conserved energy does not migrate in the form of excitons. Absorption of light results in elevation of an electron to the conductance band leaving a vacancy in the valence band. This vacancy can be filled either by back recombination of the electron from the conductance band (unproductive decay) or by transfer of an electron from the film into silicon (reaction 1, Fig. 4.12) since the affinity of this vacancy for an electron is apparently high enough to oxidize a molecule of NTCDI. As a result, similarly to the photochemistry of dye-sensitized cells [33, 34], an extra electron remains in the conductance band of silicon, while the oxidized molecules of NTCDI become cation radicals (Fig. 4.12). NTCDA molecules could not be oxidized at normal conditions [35, 36]. It is also known that OFET based on NTCDA molecules shows mobility of $10^{-4}$ cm$^2$ V$^{-1}$ s$^{-1}$ [37, 38], while NTCDI-based OFET routinely demonstrates mobilities, which are higher by factor $10^3$ [39–45].

### *4.3.5 Device Application Scope of MLE Technology*

MLE technology was developed in the end of 90th. The initial studies have been concentrated on fabrication of artificial organic superlattices and hybrid heterostructures. The first generation of MLE devices includes MLE-derived organic light-emitting diodes (see Fig. 4.7) and field-effect transistors (see Fig. 4.9). Future development of MLE technology is anticipated along the following directions: artificial photosynthesis (see Fig. 4.12), bi-interfacing, molecular spintronic (see more details on this topic in chapter "Prospects for application of GaAs doped with transition metals as a materials for spintronics" of this book), and nano-lithography technology. Most recent development in MLE devices includes studies toward fabrication of organic NLO devices (see Fig. 4.11), nano-catalytic structures. Figure 4.13 summarizes schematically directions of MLE research development and its device application.

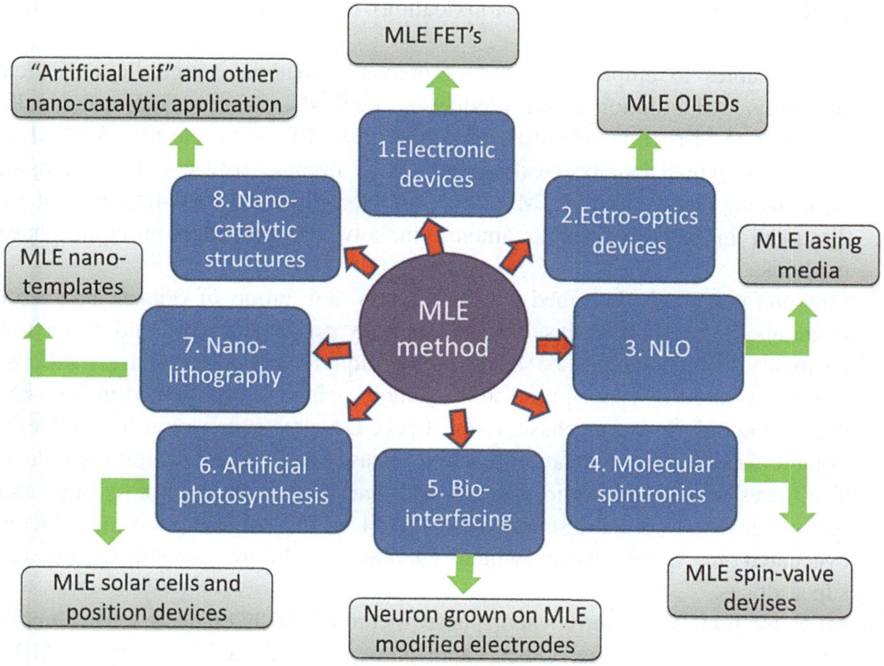


**Fig. 4.13** Application scope of MLE technology. Research directions are shown in blue boxes and device applications are shown in gray boxes

## 4.4 Distinguished Features of MLE Method and Outlook Toward Hybrid Structures

It is instructive to review the characteristic features of MLE technology in respect to other organic thin film technologies to show the potential of the MLE method in molecular nanoelectronics and review its potential directions of development.

### *4.4.1 Distinguished Features of MLE Method*

We summarize the distinguishing features of MLE method with competing technologies in deposition of organic thin films as follows:

1. MLE is a solvent-free technology. Using of solvents in regular molecular nanoelectronic devices results in decreasing of electronic properties, including charge mobility. This usually is associated with trapping charges on traces of solvent molecules, which remain embedded in matrix of useful electronics material. In addition, any traces of solvents contribute to graduate degradation of useful properties (conductivity, charge mobility, etc.) of molecular electronic device along with degradation of whole device, especially when large operation fields are applied (e.g., due to photooxidation). MLE process is 100% solvent-free technology.
2. MLE enables to employ all kind of precursor materials, regardless of their solubility. All solution-based production methods (solution molecular self-assembly (SAM) films, electrostatic SAM films, LB films, etc.) are limited to molecules, which can be easily dissolved in organic solvents. For example, NTCDA molecule, used in MLE process (see Figs. 4.4 and 4.5), cannot be dissolved in any appreciable amount in solvents for usage in liquid-phase methods.
3. Previous paragraph described MLE to assess new group of organic precursor materials. In addition to this, MLE technology provides a new kind of surface chemistry, which cannot be achieved by liquid- and vapor-phase methods. Indeed, reaction of vapor-phase self-assembly, which is highlighted in Fig. 4.5, is possible only in vapor phase. Even if NTCDA precursor would be soluble in organic solvent, the imidization reaction releases water as a reaction coproduct. The traces of water in liquid phase would react with surface amino groups and prevent growth of organic superlattices (Fig. 4.5). In MLE setup the coproduct is evacuated in vapor phase, which enables the future growth of organic superlattices.
4. MLE method is the only vapor-phase method, which enables self-assembly reactions in vapor phase. Other vapor-phase methods (CVD, UHV, OMBD, etc.) are capable to form only physisorbed films, which are designating sometimes as "pseudo-epitaxy"). These films are characterized by weak intermolecular

bonds, and these films are prone to decompose with time, especially in condition of working devices.

5. While organic superlattices can be achieved by other vapor-phase technologies (e.g., OMBD [18]), the MLE packing of these superlattices is unique as well. Figure 4.4 shows, for example, growth of NTCDI-based superlattices with aliphatic hexa-amine spacers. These superlattices are oriented orthogonally to substrate. OMBD superlattices [41] produced superlattices with NTCDA molecules, which is precursor to NTCDI superlattices in MLE case. There is no example of organic superlattices, which contains lateral π-aggraded moiety produced by solution SAM methods. The capability to create molecular devices with predesigned structures is a distinguished feature of MLE method.

### *4.4.2 Outlook for the MLE Method Toward Hybrid Superlattice*

MLE method allows to create materials on the basis of inserted organic-inorganic lattices—structures in which a series of organic layers can be included in a series of inorganic layers. Such structures were never produced and were not investigated. Such hybrid materials can have fundamentally new properties, and the range of their applications is extremely wide: from broadband photodetectors, tandem solar cells, compact energy converters, and hybrid lasers to new classes of solid-state accumulators and nano-generators and to the creation of bio-interfaces. The main characteristic of the 2nd generation of MLE setup is the ability to create stable structures.

#### **Organic-Inorganic Superlattices**

One of the challenging tasks of applied solid-state physics is to develop and study hybrid structures based on organic-inorganic superlattices including AsGa, SiGe, etc., which will be mixed by layers of *n*-type organic semiconductors of naphthalene, perylenes, etc. The distinctive feature of these structures will be epitaxial heterostructures, in which organic and inorganic layers will be connected by covalent bonds, which can sustain high temperatures, electrical fields, etc. Mixed heterostructures of this type have enormous potential in electronics and optics as their properties combine features of inorganic semiconductors with organic conducting and semiconducting materials. The possibility to study mixing of molecular properties with the properties of inorganic layers should provide a new level of understanding of solid-state physics. This kind of superlattices and heterostructures can be accomplished only by MLE technology and its derivations. One of the examples of such structure is discussed in the next section.

### Hybrid Perovskite Structures

The hybrid perovskite structures nowadays are considered to be most perspective foundation for new generation of photovoltaic device, which is believed to revolutionize market of solar cells. This development of hybrid solar cell technology can be traced back to experiments with self-assembly of biomolecules on inorganic semiconductors that have been conducted in Berkeley in the 1950s. These studies eventually resulted in appearance of new kind of solar cells, known as dye-sensitizing solar cells or Gratzel cells in 1990, which were based on self-assembly of ruthenium complex on nanoparticles of titanium oxide [31]. These solar cells were the most efficient devices for next 10–15 years, till the hybrid perovskites appeared. Nowadays the hybrid perovskites overwhelm the performance of dye-sensitizing solar cells.

However the nature of electronic processes in hybrid perovskites solar cell device remains unknown, and this impedes the future progress in achieving higher efficiencies. One of the ways to clarify the physics of solar cell device based on hybrid perovskites is to produce and study 2D structures, in which the layer of hybrid perovskites cation (e.g., $PbI_2$) can be altered with the hybrid perovskite organic anion (methyl ammonia derivatives). Currently all studies of 2D structures, which contain hybrid perovskites [46], rely on liquid-phase methods. Usage of liquid-phase methods leads in residual of organic solvents along with structural defects (typical for liquid-phase methods) in these 2D hybrid perovskite structures. Application of vapor-phase MLE technology has potential to produce the cleanest and most sophisticated 2D hybrid perovskite structures without molecules of solvents and defects, which are typical for liquid-phase SAM methods. In addition, MLE technology can employ new materials in hybrid perovskite structures, which cannot be used currently due to limited solubility of precursor materials.

## 4.5 Summary

We describe in this chapter a short introduction on new MLE method and summarize the MLE technology in the form of practical tutorial. This tutorial includes demonstration of main principles of MLE technology and few synthetic routes for vapor-phase organic superlattice assembly. Special attention is paid to several fabrication aspects of MLE technology, namely, the combination of *c*-axis bonding and *in-plane* stacking of building motifs, which led to the formation of densely packed and highly regular organic multilayers. The chemical and temporal stability of these organic structures proved to be useful in device operation. Molecular layer epitaxy approaches will be especially applicable in the "heavy duty" fields of molecular electronics, optoelectronics, bioelectronics, as well as fundamental studies.

**Acknowledgments** This research was supported by The Tomsk State University competitiveness improvement program under grant 8.2.20.2018.

## References

1. Reed MA, Zhou C, Muller CJ, Burgin TP, Tour JM (1997) Science 278:252
2. Aviram A, Ratner MA (1974) Molecular Rectifiers. Chem Physics 29:277–283
3. Anelli PL, Spencer N, Stoddart JF (1991) J Am Chem Soc 113:5131–5133
4. Petta JR, Slater SK, Ralph DC (2004) Phys Rev Lett 93:136601
5. Kelley SO, Barton JK (1999) Science 283:375–381
6. Grätzel M (2003) J Photochem Photobiol C: Photochem Rev 4:145–153
7. You J, Meng L, Song T-B, Guo T-F, Yang, Y.(M.), Chang W-H, Hong Z, Chen H, Zhou H, Chen Q, Liu Y, De Marco N, Yang Y (2015) Nat Nanotechnol:1–9. https://doi.org/10.1038/nnano.2015.230
8. Schön JH., https://en.wikipedia.org/wiki/Schön_scandal
9. Emberly EG, Kirczenow G (2000) Phys Rev B 61:5740–5750
10. *Nature Nanotechnology* 8, (2013). https://doi.org/10.1038/NNANO.2013.91
11. O'Brien E, Holt ME, Thompson MK, Salay LE, Ehlinger AC, Chazin W, Barton JK (2017) Science 355:1789
12. https://www.theverge.com/2016/8/1/12340454/gsk-google-bioelectronic-medicine-company
13. https://www.wsj.com/articles/elon-musk-launches-neuralink-to-connect-brains-with-computers-1490642652
14. Kopylova TN, Nikonov SU, Telminov EN, Gadirov RM, Degtyarenko KM, Burtman V (2016) chapter 1. In: Jensen K (ed) Heterogeneous catalysts. Nova publishers, New York, pp 1–21
15. Zhang Y, Xu L, Walker WR, Tittle CM, Backhouse CJ, Pope MA (2017) J Mater Chem C 43:13
16. Ulman A (1991) An introduction to ultrathin organic films. Boston, Academic Press
17. Yao Q, Luo Z, Yuan X, Yu Y, Zhang C, Xie J, Lee JY (2014) Sci Rep 4:3848
18. Zimmerman JD, Lassiter BE, Xiao X, Sun K, Dolocan A, Gearba R, Vanden-Bout DA, Stevenson KJ, Wickramasinghe P, Thompson ME, Forrest SR (2013) ACS Nano 7:9268
19. Kiy M, Suhner U, Gamboni I, Gunter P, Biaggio I (2000) Synth Met 111:307–310
20. Tubino R, Borghesi A, Dalla BL, Destrid S, Porziod W, Sassella A (1998) Opt Mater 9:437–444
21. Puurunen RL (2014) A short history of atomic layer deposition: Tuomo Suntola's atomic layer epitaxy (essay). Chem Vap Depos 20:332–334
22. Burtman V, Zelichenok A, Yitzchaik S (1999) Angew Chem Int Eng Ed 38:2041
23. Whitesides GM, Laibinis PE (1990) Langmuir 6:87
24. Burtman V, Ofir Y, Yitzchaik S (2001) Langmuir 17:2137
25. Burtman V, Vardeny ZV (2008) Jpn J Appl Phys 47:1165–1172
26. Yitzchaik S, Marks TJ (1996) Acc Chem Res 29:197–202
27. Burtman V (2012) In: Antsyforov GI, Ivanski AF (eds) Naphthalene: structure, properties and applications, chapter 1. Nova Publishers, New York, pp 1–58
28. Zaslavsky D, Pakoulev AV, Burtman V (2004) J Phys Chem B 108:15815
29. Sze S (1985) Semiconductor devices physics and technology. Wiley, New York
30. Hagfeldt A, Gratzel M (2000) Acc Chem Res 33:269
31. Grätzel M (2001) Nature 414:338
32. Huynh WU, Dittmer JJ, Alivisatos AP (2002) Science 295:2425
33. Glazer AN (1983) Annu Rev Biochem 52:125
34. Di Bilio AJ, Crane BR, Wehbi WA, Kiser CN, Abu-Omar MM, Carlos RM, Richards JH, Winkler JR, Gray HB (2001) J Am Chem Soc 123:3181
35. Burtman V, Zelichonok A, Pakoulev AV (2011) Int J Mol Sci 12:173–225
36. Coropceanu V, Cornil J, da Silva-Filho DA, Olivier Y, Silbey R, Bredas J-L (2007) Chem Rev 107:926
37. Laquindanum JG, Katz HE, Dodabalapur A, Lovinger AJ (1996) J Am Chem Soc 118:11331
38. Tanida S, Noda K, Kawabata H, Matsushige K (2009) Thin Solid Films 518:571
39. Facchetti A (2007) Mater Today 10:28
40. Katz HE, Johnson J, Lovinger AJ, Li W (2000) J Am Chem Soc 122:7787

41. Singh TB, Erten S, Günes S, Zafer C, Turkmen G, Kuhan B, Teoman Y, Sariciftci NS, Icli S (2006) Org Electron 7:480
42. Kao CC, Lin P, Lee CC, Wang YK, Ho JC, Shen YY (2007) Appl Phys Lett 90:212101
43. Letizia JA, Facchetti A, Stern CL, Ratner MA, Marks TJ (2005) J Am Chem Soc 127:13476
44. Lee Y-L, Hsu H-L, Chen S-Y, Yew T-R (1694) J Phys Chem C 112:2008
45. Janssen D (2006) Long-term performance versus structure of polymer materials. Ph.D. thesis, Katholieke University of Leuven, Belgium
46. Du K-z, Tu Q, Zhang X, Han Q, Liu J, Zauscher S, Mitzi DB (2017) Inorg Chem 56:9291–9302

# Chapter 5
# Prospects for Application of Gallium Arsenide Doped with Transition Metals as a Material for Spintronics

**Stanislav S. Khludkov, Ilya A. Prudaev, and Oleg P. Tolbanov**

**Abstract** This chapter is a review of the literature dealing with the production and properties of ferromagnetic gallium arsenide and the possibility of using it in spintronics. A study of ferromagnetic GaAs has been under way for two decades, and this material and the structures on its basis remain of great interest nowadays. In the review, the most attention is paid to the GaMnAs and GaFeAs materials, as well as ferromagnetic metal/GaAs structures. By the present day, the GaMnAs is one of the most promising spintronics materials. Despite the great number of studies devoted to the production and investigation of this compound, there exist urgent problems of raising the operating temperature of the structures based on GaMnAs increasing, for this purpose, the Mn content in the structures. The low solubility of Mn in gallium arsenide prevents from increasing the Curie temperature. Doping above this limit results in increasing the Mn concentration in the interstice and the development of individual phases manifesting both ferromagnetic and antiferromagnetic properties. Unlike GaMnAs, in the doping of GaAs with iron, a considerable contribution of the *d*-orbitals encourages aggregation of the Fe cautions and the formation of inclusions of the condensed magnetic semiconductor. The second-phase inclusions and Fe-based microclusters were observed in GaAs grown by different techniques. The paper discusses various effects for these materials: spin injection, giant magneto-optical effect, shift of the magnetic domain walls, interlayer exchange coupling, strong magnetic anisotropy, etc. The prospects have been shown for application of GaAs doped with transition metals in spintronics. A number of devices have been made, e.g., field-effect transistors and light-emitting diodes, a light-driven microactuator, magneto-optical materials, and magnetic field sensors.

---

S. S. Khludkov (✉) · I. A. Prudaev · O. P. Tolbanov
Functional Electronics Laboratory, Tomsk State University, Tomsk, Russia
e-mail: top@mail.tsu.ru

 
S. M. Goodnick et al. (eds.), *Semiconductor Nanotechnology*, Nanostructure Science and Technology, https://doi.org/10.1007/978-3-319-91896-9_5

## 5.1 Introduction

Spin electronics (spintronics) is a new advanced field of electronics. The term "spintronics," according to [1], was introduced by S. A. Wolf in 1996. Spintronics allows using not only the electron charge but also its spin to produce functional devices [1, 2]. In the future, it can be used to significantly increase the information content and decrease the power consumption per one bit of information and design new types of detectors, and microprocessors. Spin devices could perform a number of calculations more efficiently than their charge electronic analogs. It should be noted that at the same time, the electron spin orientation in a semiconductor can be detected both optically and electrically. It is also worth noting that the progress in this field completely depends on the existence of the materials in which the electron spin can be controlled and manipulated.

In 1988, A. Fert and P. Grünberg discovered the giant magnetoresistance effect in multilayered metal ferromagnetic–paramagnetic structures, for which they were awarded the Nobel Prize [3, 4]. Their discovery was a powerful incentive for the development of spintronics that has been drawing great attention since then [1]. The devices based on the metal structures having the giant magnetoresistance effect (GMR) [4–6] and tunnel magnetoresistance (TMR) [7] have found a wide practical application.

In their review, Zakharchenya B. P. and Korenev V. L. [2] underline the importance of integrating magnetism into the semiconductor structure of modern computers. In this connection, the development of an object combining the properties of a ferromagnetic and a semiconductor seems very promising. It should be noted that in this case, it is necessary to use ferromagnetic semiconductors that, on the one hand, can be integrated with the materials which are the basis of modern electronics, i.e., Si, Ge, and GaAs, and, on the other hand, remain ferromagnetic at room temperature.

Semiconductors are given ferromagnetic properties by being doped with magnetic impurities. As magnetic impurities, transition metals are used whose atoms have inner atomic shells that are not completely filled [8]. At small concentrations, such magnetic ions will induce in a semiconductor a paramagnetic or, at large concentrations, a ferromagnetic behavior [2]. These materials were first given the name of "diluted magnetic semiconductors" (DMS) and later "diluted ferromagnetic semiconductors" (FMS).

Ferromagnetic semiconductors were known earlier. These are europium chalcogenide and cadmium–chromium spinels. When studying them, the researchers [9], for the first time, established a link between ferromagnetism and semiconductor properties. However, they have low Curie temperature and poor compatibility with the requirements of the semiconductor technology in modern electronics. The DMS investigation was started with the II–VI compounds and oxides followed by the III–V compounds: InAs, GaAs, and later by GaN. Since the 1980s, such diluted magnetic semiconductors as CdMnTe and ZnMnSe have been studied extensively [10]. Along with the bulk materials, the structures with quantum wells (QW) and superlattices containing ultrathin DMS layers were also studied.

Successful growth of Mn-doped semiconductor III–V compounds InMnAs [11] and GaMnAs [12] using the method of the low-temperature molecular beam epitaxy (LT-MBE) and the discovery of ferromagnetism induced by the current have resulted in a new stage of studying the DMS. These materials are attractive because they are compatible with the ASIC heterostructures and QWs based on the III–V compounds used in real semiconductor devices. Besides, the Mn atoms in them are acceptors and have local magnetic moments, which allow a unique possibility of studying ferromagnetism induced by the holes. Among the above materials, gallium arsenide is of the greatest interest for the following reasons. Firstly, weakly doped GaAs (with a carrier concentration less than $2{\cdot}10^{16}$ $cm^{-3}$) is interesting for studying spin phenomena, since at temperatures below 5 K, the spin relaxation time in it is large (100–300 ns) [3]. Secondly, when doped with transition metals, it can retain ferromagnetism at room temperature.

For quite a long time, since the late 1990s, GaMnAs has been the prototype of the DMS, so lots of experimental and theoretical effort have been devoted to studying this material [13].

## 5.2 Gallium Arsenide Doped with Manganese

GaMnAs films possessing ferromagnetic properties were first produced by the Ohio group in 1996 [12]. The possibilities of producing ferromagnetic ordering in GaMnAs are limited by the weak solubility of the Mn impurity in GaAs (about 0.1%). However, the above limitation was overcome by the use of the thermodynamically nonequilibrium method, i.e., the low-temperature molecular beam epitaxy. This method was used to produce GaMnAs layers with the Mn content considerably exceeding its solubility in GaAs (up to 10%). Since then, GaMnAs has been widely studied by many research teams both as a model DMS and from the practical viewpoint of using it as a spintronics material. Various techniques of producing the material and a wide spectrum of investigation methods have been applied. The main task was to clarify the nature of ferromagnetism in the above material and produce samples with the Curie temperature ($T_c$) above room temperature. It should be borne in mind that $T_c$ is the temperature above which magnetization of a ferromagnetic disappears and it becomes a paramagnetic.

In a number of works [14–18], first-principles calculations were made, and simulation of various DMS properties of GaMnAs was performed. Numerical simulation has allowed establishing the nature of the magnetic anisotropy of the compound [14] and explaining the magnitude of the orbital and spin magnetization in the above material [15]. An exact expression for the magnetic circular dichroism of GaMnAs has been derived [16]. The calculation of a magnetic topological semiconductor, which is GaMnAs with the Mn-based magnetic inclusions, has been made in [17]. The effect of the hydrostatic pressure on the Curie temperature of GaMnAs has been studied. According to the first-principles calculation data [18], at normal pressure for GaMnAs with the Mn content $x = 12.5\%$, it must be 181 K.

With increasing pressure up to $P = 6$ GPa, the Curie temperature also increases, with the value of $dT_c/dP$ being +4.3 and ≈ +2.2 K/ GPa for $x = 12.5\%$ and $x = 6.25\%$, respectively.

### 5.2.1 Production of GaMnAs Layers

The use of the thermodynamically nonequilibrium method of the LT-MBE has allowed producing GaMnAs layers with the Mn content considerably exceeding its solubility [8]. The low-temperature (200–300 °C) growth depresses Mn segregation and the formation of the second phase. $Ga_{1-x}Mn_xAs$ layers with the concentration of $Mn_x = 0.015–0.10$ and a thickness of 150–200 nm are usually grown on a substrate of semi-insulating gallium arsenide. The film properties significantly depend on the growth temperature and the pressure of the arsenic vapor. At the beginning, GaMnAs layers were produced at the Curie temperature about 60 K [12], and then it was increased up to 110 K [8, 19]. According to the data [20], the Curie temperature depends on the composition of $Ga_{1-x}Mn_xAs$: it grows up to 110 K in proportion to $x$ (up to $x = 0.053$) and then drops. It should be noted that the hole concentration was only 15% of the Mn atom concentration, which may be caused by the compensation of the Mn acceptor atoms with deep donors.

Further success in raising the Curie temperature was achieved due to the use of annealing after the growth of $Ga_{1-x}Mn_xAs$ films [21–23]. Low-temperature annealing of the films leads to an increase in $T_c$ and the hole concentration. The $Ga_{1-x}Mn_xAs$ films grown by the LT-MBE technique and annealed at temperatures of 220–310 °C had the ferromagnetic properties within the entire $x$ interval and the Curie temperature up to 160 K [23]. Later, $Ga_{1-x}Mn_xAs$ films were produced with the Curie temperature of 185 K [24, 25] and then with $T_c \approx 200$ K [26, 27], which, according to the authors, was the best result. Thus, due to the improved technology of producing $Ga_{1-x}Mn_xAs$ films, the Curie temperature rose from 60 to 185–200 K.

Along with the MBE technique, GaMnAs diluted magnetic semiconductors were also produced using the method of implantation of manganese into semi-insulating GaAs [28]. The implantation dose of Mn was $2 \cdot 10^{16}$ cm$^{-2}$ (4% atomic) and the ion energy of 100, 200, and 300 keV. Arsenic was implanted simultaneously with Mn with an ion energy of 270 keV and an implantation dose of $2 \cdot 10^{16}$ cm$^{-2}$, annealing of the samples was performed at 750°C. The advantage of the above method of doping is the production of the DMS with $T_c$ above room temperature.

Depending on the doping conditions, the Curie temperature was varied within 352–388 K. In all samples the hysteresis on the magnetization curves was observed at room temperature with the coercive fields being 50 Oe. It is worth noting that a coercive field is the magnetic field strength at which the substance magnetization becomes zero. It is considered that ferromagnetism of GaMnAs doped with manganese during implantation is caused by the incorporation of the second phase, i.e., by nanoclusters. Nanoclusters are also formed in $Ga_{1-x}Mn_xAs$ produced by the LT-MBE method, if the films after the growth are subjected to annealing at a high temperature. Thus, in Ref. [29], the nanoclusters with the Curie temperature above room temperature were observed in the $Ga_{1-x}Mn_xAs$ films grown by the LT-MBE at 280 °C and annealed at 600 °C. In GaMnAs doped with Mn during ion introduction or the MOCVD using a laser evaporator, $T_c$ reaches 400 K, which is explained by the formation of the MnAs and MnGa clusters during annealing [30, 31]. It has been established that the magnetic properties of the GaMnAs layers are affected by the elastic stresses (compression, tension) as a result of the growth on the GaAs substrate of the buffer layers of $In_xGa_{1-x}As$ or $In_xGa_{1-x}P$ differing from GaAs in the crystal lattice parameter [31].

To produce $Ga_{1-x}Mn_xAs$ DMS, the method of high-frequency magnetron sputtering was also used. The polycrystalline layers of $Ga_{1-x}Mn_xAs$ with a different Mn concentration were produced and investigated in [32, 33]. Manganese can enter the polycrystalline layers at larger concentrations than in the monocrystalline material, with crystallites formed of the size of several nanometers. The layer properties change during a subsequent treatment in hydrogen, e.g., its conductivity decreases.

Along with thermal annealing, pulsed laser annealing was also used to produce ferromagnetic GaMnAs [34].

In [35, 36], the LT-MBE technique was used to obtain (Ga,Mn)As thin layers with a thickness of 25 nm and the Mn concentration of 7%. $T_c$ in the above layers after the growth was $\approx$ 70 K, which after annealing at 190 °C increased up to $\approx$ 140 K. At the Mn concentration of 2% in $Ga_{1-x}Mn_xAs$, the metal-insulator transition was found. It was shown that there was a significant difference in the behavior of the polarized holes in the weakly doped semi-insulating and heavily doped metal layers.

In the above works, the GaMnAs epitaxial layers were produced, as a rule, on GaAs semi-insulating substrates. Later, the authors in [37] demonstrated the possibility of growing GaMnAs layers with the Mn content of 6.3% by the MBE method on Si substrates. It should be noted that in most studies, the presence of ferromagnetism was determined by measuring the dependence of magnetization on the value and the polarity of the magnetic field by the formation of the hysteresis loop. The authors in Refs. [38, 39] have developed a whole complex of the methods of investigation of GaMnAs thin epitaxial films, which allows the production of the magnetostatic and dynamic film parameters.

### *5.2.2 Dependence of the Curie Temperature on the Manganese Concentration in GaMnAs*

To establish the mechanism of change of the Curie temperature, the authors in [40] have conducted a detailed experimental and theoretical investigation of GaMnAs. They examined GaMnAs films with a thickness of 25 and 50 nm and with a low and high Mn concentration grown at the temperatures of 300 and 180°C, respectively. After the growth, the samples were subjected to annealing at 190°C. The Mn concentration was determined with the help of the secondary-ion mass spectrometry

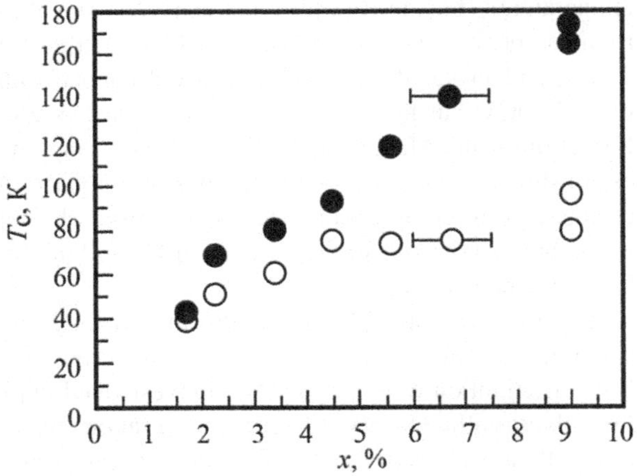


**Fig. 5.1** Dependence of the Curie temperature vs. the full manganese concentration. The hollow symbols – immediately after the growth, the painted ones – after annealing [40]

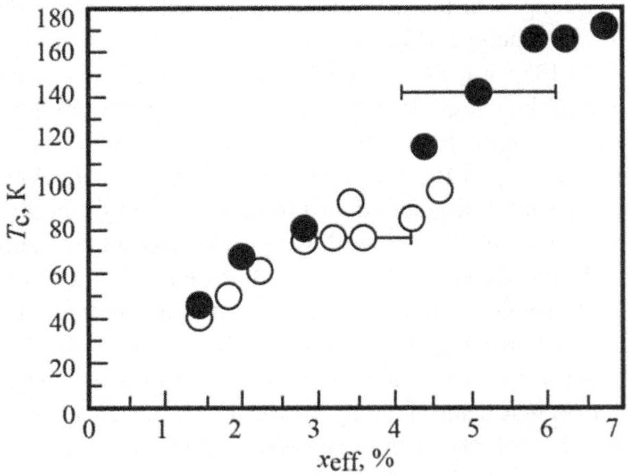


**Fig. 5.2** Dependence of the Curie temperature vs. the concentration of the Mn uncompensated acceptor atoms at the Ga lattice points – $x_{eff}$ [40]. The hollow symbols – immediately after the growth, the painted symbols – after annealing [40]

on the control samples with a layer thickness of 1 μm. The hole concentration and the Curie temperature ($T_c$) were found by means of a single measurement of the magnetoresistance and the Hall effect. The dependence of the Curie temperature on the Mn concentration is shown in Fig. 5.1.

One can see that Mn concentration increases from 1.7 to 9% and the $T_c$ of the unannealed samples increases from 40 to 90 K. After annealing, $T_c$ of the samples increases considerably reaching 160–170 K. Figure 5.2 presents the dependence of the Curie temperature on the concentration of the uncompensated acceptor centers of Mn($x_{eff}$):

$$x_{eff} = x_s - x_i,$$

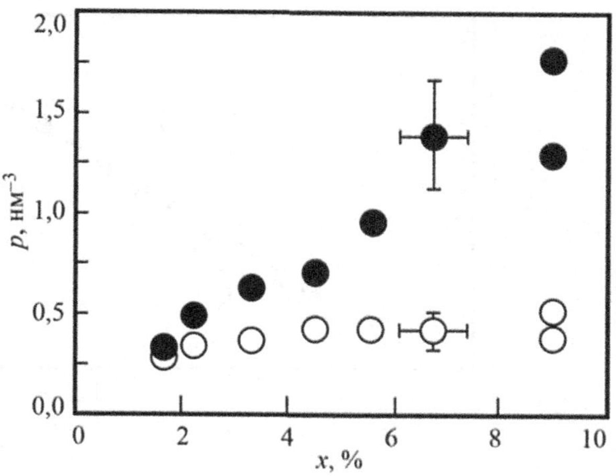


**Fig. 5.3** Dependence of the hole concentration vs. the full Mn concentration [40]. The hollow symbols – immediately after the growth, the painted ones – after annealing

where $x_s$ and $x_i$ are the concentrations of the Mn atoms at the gallium lattice points ($Mn_{Ga}$) and the interstice ($Mn_i$). From Fig. 5.2, it follows that with increasing $x_{eff}$, $T_c$ grows, practically, linearly.

Figure 5.3 shows the dependence of the hole concentration on the Mn concentration measured by the Hall method.

The experimental results obtained in the above study agree well with the theoretical calculations. It has allowed the authors to make a conclusion that in high-quality GaMnAs samples, the Curie temperature increases linearly with the concentration of the uncompensated Mn atoms without any signs of saturation. At the concentration $x_{eff}$ equal to 10%, ferromagnetism can be observed at room temperature.

The data on the magnetotransport and magnetization agree with the model, according to which the Mn impurity atoms getting into the interstice during the growth act like double donors and compensate the adjacent acceptor centers of $Mn_{Ga}$ as a result of the strong antiferromagnetic $Mn_{Ga}$–$Mn_i$ coupling. However, these defects can be efficiently eliminated during the aftergrowth annealing.

So, if at a full concentration of Mn equal to 9% immediately after the growth the value of $x_{eff}$ is 4.6%, then after annealing, it increases up to 6.8%. Thus, the analysis performed in this work has allowed the authors to make a conclusion that in high-quality $Ga_{1-x}Mn_xAs$, there is no fundamental obstacle to achieving a high concentration of the uncompensated $Mn_{Ga}$ atoms and producing ferromagnetism at room temperature.

### *5.2.3 Dependence of the Curie Temperature on the Hole Concentration in the Layers of GaMnAs*

In a number of experimental works, the authors investigate the dependence of the Curie temperature on the hole concentration ($p$) in GaMnAs layers and make comparison with the theory based on the Zener kinetic $p$–$d$ model. According to

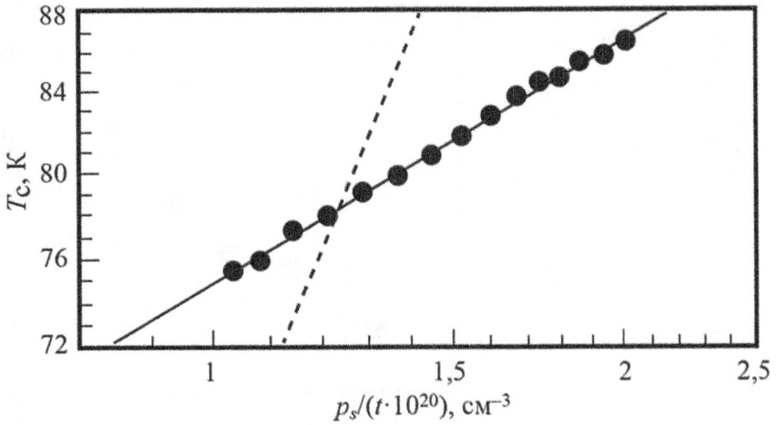


**Fig. 5.4** Curie temperature ($T_c$) vs. the hole concentration (layered $p_s$ concentration divided by the layer thickness $t$). The circles (experimental data) and the dashed line (the calculation) by the Zener kinetic model [43]

the model, ferromagnetism is caused by the interaction between the free charge carriers and the spins of Mn ions. The use of the above model to predict $T_c$ in various semiconductors has allowed the authors [41] to make a conclusion that GaMnAs is the most promising material.

In Refs. [42–45], the experimental data on the dependence of the Curie temperature on the hole concentration $p$ were obtained from the measurements of the anomalous Hall effect on the structures similar to field transistors (the FET-type structure) with a channel from GaMnAs. In the above structures, the hole concentration in the GaMnAs layers changes due to the applied external voltage. Figure 5.4 shows the dependence of $T_c(p)$ based on the data in [43].

One can see that $T_c$ increases with increasing hole concentration according to the power law with the power index $\gamma \approx 0.2$. In accordance with the Zener theory, the coefficient $\gamma$ is 0.7. Assuming that the discrepancy of the values of $\gamma$ is related to the inaccuracy of determination of the hole concentration as a result of their nonuniform distribution in the layer, the authors think that the theory and experiment are in a satisfactory agreement.

Similar investigations conducted in [46–48] have shown that to achieve a considerable effect in raising $T_c$, GaMnAs with the Mn content over 1% should be used as a channel in the FET-type structure. Besides, the effect value depends on the channel thickness. A decrease in the thickness of the GaMnAs film serving as a FET channel increases the effect. For instance, according to [47], a decrease in the channel thickness from 7 to 3 nm has resulted in a threefold increase in $T_c$. In a FET structure, in which the GaMnAs film was used with the Mn content of 6% and a thickness of 7 nm, a strong exchange interaction was observed [48]. It is also noted that in the above structures, a ferroelectric polymer is successfully used as a gate [46, 47]. It should be also underlined that the FET-type structure with a GaMnAs

DMS as a channel draws great attention of the researchers because it can find practical application, as it allows electrical control of the DMS magnetic properties.

However, as it is noted in [42], in spite of the active experimental and theoretical studies and even a successful demonstration of the prototypes of spin devices, no deep insight into the physics of the phenomenon of the $Ga_{1-x}Mn_xAs$ DMS has been achieved yet. In most proposed models, the holes produced by the Mn impurity are assumed to play the key role in the interaction with the Mn ion spins.

### *5.2.4 Magnetotransport Properties of GaMnAs*

In a detailed study of various properties of the GaMnAs DMS, a number of new interesting phenomena have been found. One of them is the shift of the magnetic domain walls (MDW). Domains are the regions of uniform magnetization in a ferromagnetic whose size can change due to the domain wall movement. Usually, it happens under the influence of the external magnetic field. In GaMnAs films, there occurred a shift of the magnetic domain walls as a result of the injection of the charge carriers [42]. The MDW shift in GaMnAs at 60 K took place at a current density of $10^5$A/cm$^2$, which is much lower than in metal ferromagnetics. For instance, in the ferromagnetic NiFe nanowires, the MDW shift occurs at densities of $10^7$–$10^8$A/cm$^2$. It should be noted that, at the same time, in the GaMnAs films, the speed of the domain movement caused by the control current and found with the help of the Kerr effect was changed by more than five orders of magnitude (up to 20 m/s). Thus, it was shown that control of the magnetization vector of GaMnAs could be achieved only by an electric field without application of external magnetic fields. The electric field control of magnetization without the use of external magnetic fields is an outstanding physical phenomenon that is also important in the practical aspect of reducing power under reversible magnetization in high-density magnetic memory cells.

In Ref. [49], the authors studied the X-ray magnetic circular dichroism in the Faraday configuration in the ferromagnetic $Ga_{1-x}Mn_xAs$ DMS with $x = 0.025$ and $x = 0.01$. The absorption spectra were measured within the energy interval of 636–648 eV, with the circular light polarization exceeding 95%. They investigated GaMnAs films with a thickness of 500 nm and a hole concentration of $10^{18}$ cm$^{-3}$ at a temperature of 40 K. The parameters of the Mn electron structure in GaMnAs were obtained. They also established the *p–d* hybridization of the As–Mn link and the constant of the weak *p–d* exchange.

A strong magnetic anisotropy was found in GaMnAs films which determine the magnetization vector. It essentially depends on the mechanical stresses (tension or compression) in the film caused by the discrepancy of the lattice parameters of the film and the substrate as well as on the hole concentration. Varying the type of the mechanical stresses, it is possible to change the magnetization vector. In Ref. [50], the magnetic anisotropy in the film was produced by the incorporation of thin layers of InGaAs. In the GaMnAs samples with the Mn content of 3%, the magnetization vector was perpendicular to the sample plane, and the magnetization saturation was 34 kA/m [51]. The investigations have shown that the magnetic anisotropy can be varied through changing the hole concentration by doping with beryllium [52] or by hydrogenation of the samples [53].

In the review [54], the authors have systematized the results on the magnetic anisotropy of GaMnAs films. They consider memory devices, in which information is recorded via changing the magnetization vector, to be one of the most probable fields of application of GaMnAs in spintronics. Therefore, in the practical aspect, magnetic anisotropy is one of the most significant physical phenomena.

In Ref. [55], the authors developed a method of measurement of the magnetic homogeneity of GaMnAs using the frequency-dependent ferromagnetic resonance having a high spatial resolution. Using the above method, they found in GaMnAs films the magnetic inhomogeneities both of the submicron and submillimeter size. Due to the ferromagnetic nanoclusters of MnAs incorporated in GaAs, a giant magneto-optical effect was observed in GaMnAs [56], with the Faraday rotation angle at a wavelength of 0.98 μm being about 0.2°/μm at the magnetic field strength of 2 kOe. It was also shown that the presence of the MnAs clusters in GaMnAs could lead to a shift of the radiation maxima in the photoluminescence spectra [57, 58].

In Ref. [59], the authors developed methods of investigation of the fundamental optical properties of the GaMnAs DMS thin films with a low (1–2%) and high (4–6%) Mn content using the photoreflection spectroscopy, the Raman spectroscopy, and the X-ray high-resolution diffractometry.

Thus, by the present day, the GaMnAs DMS is one of the most promising spintronics materials. Despite the great number of studies devoted to the production and investigation of this compound, there exist urgent problems of raising the operating temperature of the structures based on GaMnAs increasing, for this purpose, the Mn content in the structures. The low solubility of Mn in gallium arsenide prevents from increasing the Curie temperature. Doping above this limit results in increasing the Mn concentration in the interstice and the development of individual phases manifesting both ferromagnetic and antiferromagnetic properties, which essentially changes the characteristics of these systems.

## 5.3 Gallium Arsenide Doped with Iron

Doping of gallium arsenide with iron was commonly used to produce a semi-insulating material and only recently to produce a ferromagnetic material. It can be implemented by various methods: in the course of the growth of monocrystals and epitaxial layers, by diffusion, or ion implantation. In the present section, we have systematized the data on the electric, structural, and magnetic properties of gallium arsenide doped with iron produced by various methods. It should be noted that unlike GaAs/Mn, in the doping of GaAs with iron, a considerable contribution of the *d*-orbitals encourages aggregation of the Fe cautions and the formation of inclusions of the condensed magnetic semiconductor [45]. The second-phase inclusions and Fe-based microclusters were observed in GaAs grown by the Czochralski method in the layers produced by the methods of epitaxy, implantation, and diffusion.

### 5.3.1 GaAs Doped with Iron During the Growth of Monocrystals

GaAs has been doped with iron in the course of the growth of monocrystals since 1965 [60]. It has been shown that in GaAs crystals grown by the Czochralski method, the Fe impurity is the acceptor and produces a level spaced at 0.52 eV from the valence band [60, 61]. During the growth of GaAs crystals by the Czochralski flux method, the conditions were created at which the Fe concentration exceeded its solubility in GaAs [62].

It resulted in the formation of the iron-based precipitates whose composition was identified as the compound $Fe_3GaAs$. Measurement of the magnetic properties of the precipitates showed them to possess the ferromagnetic properties with the Curie temperature about 100 °C.

In Refs. [63–65], the authors studied the magnetic properties of the GaAs/Fe crystals produced by the method of directional crystallization. The iron concentration changed from $1 \cdot 10^{18}$ up to $6 \cdot 10^{20}$ $cm^{-3}$ by means of varying the growth conditions. In the above crystals, irrespective of the Fe concentration, one could observe the magnetic phase transition with the Curie temperature $T_c = 460$ K and large values of susceptibility. Investigation of the magnetic phase transitions in the Fe–Ga and Fe–As systems [63] has shown that among various compounds on their basis, $Fe_8Ga_{11}$ with $T_c = 750$ K and FeAs with $T_c = 130$ K possess ferromagnetic properties, and $Fe_2As$ has antiferromagnetic properties with the Néel temperature $T_N = 850$ K. When investigating the GaAs/Fe crystals heavily doped with iron using the method of the electron paramagnetic resonance, the authors [65] came to the conclusion that the above crystals could be considered as a two-phase system. One phase is supermagnetic regions with a pronounced correlation in the Fe atom disposition with the Curie temperature $T_c = 460$ K, while the other phase is a disordered magnetic system of the Fe atoms in GaAs between the supermagnetic regions.

### 5.3.2 GaAs Doped with Iron During Epitaxy

Doping of GaAs with iron during vapor-phase epitaxy was performed at a growth temperature of 750 °C [66–68], with an iron concentration of ~$2 \cdot 10^{17}$ $cm^{-3}$ according to the data of the secondary-ion mass spectrometry (SIMS), which corresponds to Fe solubility at the given temperature. To compensate the uncontrolled acceptor impurities, the donor impurity of sulfur and tin was introduced simultaneously with iron. As a result, the authors produced GaAs semi-insulating layers with a resistivity of up to $2 \cdot 10^5$ Ohm·cm and a hole mobility of 300–400 $cm^2/V \cdot s$. The iron activation energy was 0.5 eV. In the layers heavily doped with iron, the second-phase inclusions containing Fe, FeAs, and $FeAs_2$ were found by the methods of electron microscopy and electron diffraction.

In GaAs doped with iron during hydride vapor-phase epitaxy at a growth temperature of 735 °C, the iron concentration was varied from $4 \cdot 10^{16}$ to $4.5 \cdot 10^{20}$ $cm^{-3}$ [69]. In the samples with a high concentration of Fe, it was possible to observe the acceptor Fe center with an activation energy of 0.5 eV. The high ohmic layers were produced at the minimum Fe concentration. It should be noted that, according to the authors, the high resistivity was caused by the EL2 centers of the donor type with an activation energy of 0.72 eV.

To produce GaAs/Fe possessing ferromagnetic properties, doping of GaAs is commonly performed during the nonequilibrium process of the low-temperature vapor-phase epitaxy at temperatures of 260–350 °C followed by annealing at 400–600 °C [70–72]. It should be noted that GaAs/Fe retains its semi-insulating properties, practically, at all degrees of doping with iron (at an iron concentration from $1 \cdot 10^{18}$ to $1.5 \cdot 10^{21}$ $cm^{-3}$) [70]. The layers produced by the LT-MBE were of two types: (*1*) the layers possessing the properties of a diluted magnetic semiconductor (DMS) GaFeAs [70, 71] and (*2*) GaAs/Fe layers with the second-phase incorporations [72]. GaFeAs (DMS) layers with a thickness of 700 nm were produced at substrate temperatures of 260 and 350 °C. The Fe concentration was $1.5 \cdot 10^{21}$ $cm^{-3}$ ($x = 0.07$). The layers were investigated by the methods of the electron diffraction, atomic force and transmission microscopy, and SIMS. It was established that at low growth temperatures, the Fe magnetic impurity was uniformly distributed in the GaAs matrix. A similar result for the GaFeAs films that had not been annealed was obtained in Ref. [73]. Subsequent annealing at a temperature of 580 °C leads to the formation of the FeAs complexes and Fe-Fe clusters. At long annealing, there occurs the formation of the $FeAs_2$ phase [73]. In Ref. [74], the authors produced GaAs/Fe layers which possessed the paramagnetic properties immediately after the growth but, however, after annealing at 600 °C acquired the ferromagnetic properties as a result of the formation of the Fe-based ferromagnetic clusters.

According to the triple-phase diagram of Ga–Fe–As, the following compounds can form in this system: FeAs, $FeAs_2$, $Fe_2As$, $Fe_3Ga$, $Fe_3Ga_4$, and $Fe_3Ga_{2-x}As_x$. They all possess various magnetic properties, e.g., $FeAs_2$ is a diamagnetic, $Fe_3Ga$ and $Fe_3Ga_{2-x}As_x$ are ferromagnetics, FeAs and $Fe_2As$ are antiferromagnetics, and $Fe_3Ga_4$ is a meta-magnetic [75, 76].

In GaAs/Fe structures, a unique phenomenon has been found and, namely, the photomagnetic effect of controlling magnetization under irradiation [72, 77].

This effect was first observed in InMnAs films [78]. The authors [72] described the phenomenon of magnetism induced by the current at room temperature in the composite structures of GaAs/Fe obtained in the following way. Three-dimensional Fe grains were produced by the LT-MBE at a temperature of 580 °C followed by a GaAs layer grown on them. The above procedure was repeated several times. As a result, they obtained a multilayer composite structure consisting of GaAs and $(GaAs)_m(Fe)_n$ of a different composition. In the above structure, Fe clusters were formed with a diameter of 7 Å possessing the magnetic moment up to 34 $\mu_B$. It was these clusters that were responsible for the magnetization enhanced by light.

Hybrid GaAs/Fe structures have also been produced in [79]. At a large content of Fe, they manifest a wide negative peak with the Kerr rotation of 0.2° at a photon energy of 0.9 eV over the whole temperature range. The spectral dependence is very similar to that for pure Fe.

The authors in [80, 81] have produced GaFeAs layers in which during the growth and annealing there form inclusions of the second phase of the $Fe_3Ga_4$ and $Fe_3Ga_{2-x}As_x$ compositions. In their opinion, the ferromagnetic $Fe_3Ga_{2-x}As_x$ and its derivatives are a stable phase of the triple system of Ga–Fe–As, and the meta-magnetic $Fe_3Ga_4$ is a metastable phase, though it can play a key role in the devices capable of operating at room temperature. The GaFeAs layer obtained in that way possesses the effect of photo-accelerated magnetization at room temperature.

### *5.3.3 GaAs Doped with Iron During Implantation and Diffusion*

In doping of GaAs with iron during implantation and subsequent annealing, one could observe the formation of the precipitates based on Fe of the composition of $Fe_3GaAs$ or $Fe_3Ga_{2-x}As_x$ with a size from units to hundreds of nanometers [82, 83].

Gallium arsenide doped with iron during diffusion was investigated in [84–86]. The diffusion of iron was performed at a temperature of 1100 °C, and then the samples were annealed at 900 °C for 0.25–3 hrs. According to the data of measurement of the temperature dependence of the electric conductivity and the Hall constant, the authors established the position of the acceptor level produced by iron, which was (0.53 ± 0.01) eV relative to the valence band maximum [84].

It was also found that the iron center concentration determining the hole concentration and the impurity activation energy of Fe in GaAs did not depend on the thermal treatment regime. The temperature dependence of the hole mobility in GaAs/Fe is well described by the expression $\mu \sim T^{-3/2}$, which suggests the predominance of the charge carrier scattering on the phonons.

Two methods were used to conduct structural investigations: atomic force microscopy (AFM) and transmission electron microscopy (TEM). AFM studies were performed on the sample chips across the cleavage plane. The samples for TEM were prepared in the form of thin foils parallel to the plane (100). Measurements of the relief topology showed that on the chip surface of the samples subjected to additional annealing at 900 °C, in the background of the smooth or step-shaped surface, one could observe local heterogeneities with a diameter of 50–500 nm and a height of 1.5–50 nm. A typical pattern is shown in Fig. 5.5. One can see distinct

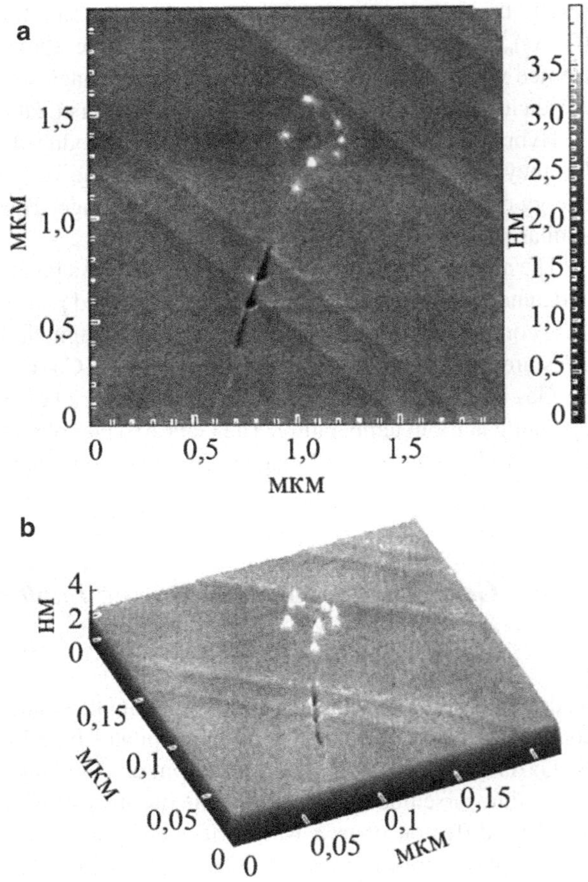


**Fig. 5.5** AFM image of the chip surface relief of GaAs/Fe: (**a**) two-dimensional; (**b**) three-dimensional

heterogeneities of the relief in the form of aggregates of hills with a diameter of ~60 nm and a height of ~2 nm.

The observed heterogeneities are the second-phase inclusions formed as a result of decomposition of the supersaturated solid solution of GaAs/Fe during the annealing at 900 °C. From the AFM, it follows that the inclusions, as a rule, are distributed nonuniformly. Sometimes on the chip surface, there are larger inclusions (up to 1 μm). Figure 5.6a shows AFM images of two inclusions with a diameter ~0.4 μm. The AFM image of the inclusions shown in Fig. 5.6b indicates that they possess the ferromagnetic properties at room temperature.

The second-phase inclusions of a smaller size are observed on the samples subjected to rapid cooling after diffusion. Apparently, their emergence is caused by a rate of cooling which is not high enough for quenching.

TEM investigations have shown that after annealing, there appear platelike inclusions in the material in the form close to a disk. The inclusion diameter is 100–600 nm and a thickness of tens of nanometers. The inclusions are located in parallel to the planes {100}.

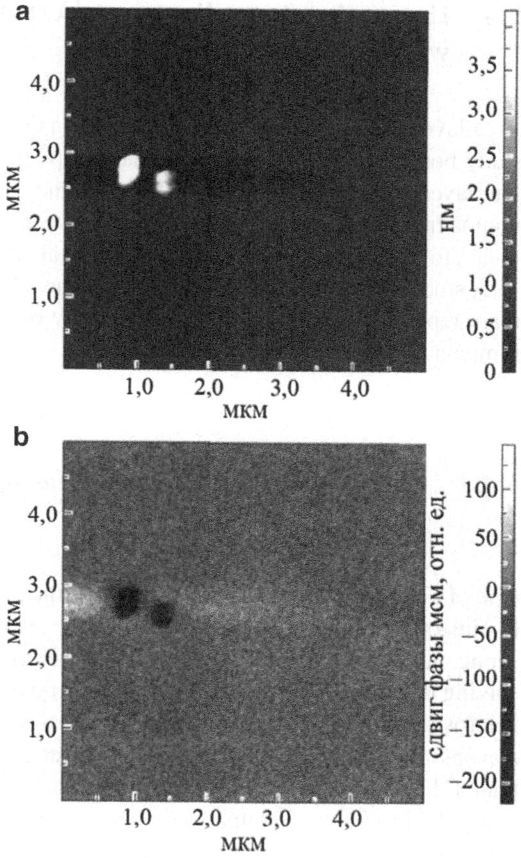


**Fig. 5.6** Image of the chip surface of GaAs/Fe: (**a**) the AFM topography; (**b**) the AFM image

The moiré patterns and the diffractions of the fast electrons indicate the ordered lattice of the second phase, with the crystal structure possessing the cubic symmetry. Evaluation shows that the concentration of the second-phase particles is $\approx 5{\cdot}10^{10}$–$5{\cdot}10^{11}$ cm$^{-3}$.

Thus, in gallium arsenide doped with iron during diffusion at the temperature $T_d = 1100$ °C corresponding to the maximum solubility of Fe in GaAs, there occurs disintegration of the oversaturated solid solution with the formation of the second-phase inclusions during annealing at a temperature by 200 °C lower than $T_d$. The inclusions have a disklike form with a diameter of 50–1000 nm and a thickness of 1.5–50 nm and possess the ferromagnetic properties at room temperature.

From the above results, it follows that in GaAs/Fe at all methods of doping (MBE followed by annealing, implantation of the Fe ions, iron diffusion), there appear inclusions of the second phase, predominantly, of $Fe_3Ga_4$ and $Fe_3Ga_{2-x}As_x$ of a size from units to 1000 nm. Due to the Fe-based precipitates, GaAs doped with iron, irrespective of the way of production, possesses the ferromagnetic properties at room temperature, which is of great interest for spintronics.

## 5.4 Heterostructures Based on Ferromagnetic GaAs Doped with Transition Metals

Two-layer and multilayer structures based on GaMnAs and GaFeAs are under active study both from the physical and practical points of view. The phenomenon of the interlayer exchange coupling (IEC) of the magnetic layers separated by the nonmagnetic medium is most widely investigated. This phenomenon attracted great attention after its discovery in the metal structures of Fe/Cr/Fe [87].

A study of the phenomenon in the structures based on the DMS thin films, whose main representative is GaMnAs, opens new possibilities of investigating magnetic semiconductors.

### *5.4.1 Production and the Properties of the Fe Films Based on GaAs*

One of the important tasks of spintronics is the production of spin-oriented electrons for injection into a semiconductor (further referred to as "spin injection") on the semiconductor surface of a ferromagnetic electrode [88]. A promising method of solving this problem is the growth of monocrystal Fe films on the GaAs surface. The heterostructure of Fe/GaAs attracts the researchers' attention for a number of reasons. Firstly, Fe and GaAs slightly differ in the crystal lattice parameters (less than 1.4%), which ensures a high crystal perfection of the grown structures. Secondly, Fe is an ideal ferromagnetic material with a high Curie temperature (768 °C) and a high value of magnetic saturation [89–91]. Fe epitaxial layers on GaAs substrates have been studied since the 1980s [89, 90].

Monocrystal films of Fe and ferromagnetic compounds on its basis on GaAs substrates are grown by the MBE. To decrease the interaction between Fe and GaAs, either the low-temperature growth (at 150 °C) was used or an ultrathin Al layer (two monolayers) was put between Fe and GaAs [92, 93]. The Fe layers were investigated by scanning-tunneling microscopy. The poly- and monocrystal layers of either the FeAs composition or that of $FeAs_2$ with a thickness of 100–585 nm as well as the layers of $Fe_3GaAs$ with a thickness of 25–80 nm possessed good electric and magnetic properties and, according to the authors, were suitable for the production of magneto-electronic devices [88, 94].

In [95] the authors obtained and studied ultrathin Fe layers of various thickness (2.5–140 monolayers) on the GaAs surface. The layers consisting of 3.5 monolayers possessed the ferromagnetic properties at room temperature; however, the spin injection efficiency in the above structures was small (about 5–6%). Ultrathin epitaxial two-layer films of Fe/MgO were grown on the (Ga, Mn)As surface. As it follows from the measurement of the current through the barrier, the carrier tunnel transport takes place in the above films [96].

In studying various properties of the ferromagnetic films on semiconductor substrates, great attention is paid to the investigation of magnetic anisotropy, which is important from the physical point of view and its application in spintronic devices. Investigation of the films of the $Fe_{31}Co_{69}$ composition on the GaAs substrates has shown that uniaxial magnetic anisotropy (UMA) considerably increases during aftergrowth annealing as a result of the interface structural modification [97].

According to the data [91], the Fe films of the cubic modification grown by the MBE on the GaAs surface are characterized by a very strong anisotropy in the plane of the film. At the same time, the controlled reconstruction of the substrate during the growth of the Fe film can allow manipulation of the anisotropy of the Fe films. The magnetization vector of a thin Fe film on GaAs can be completely reversed without applying an external magnetic field by a slight change of the temperature by several degrees near room temperature in the thermal cycle [98]. It was implemented on the samples in which the Fe layer with a thickness of 5 nm on GaAs was covered with a protective layer of Au with a thickness of 5 nm.

Investigation of the structures of Fe/GaAs and CoFe/GaAs has shown the spin injection to depend significantly on the conditions of the film production [96, 99, 100]. The $Co_{70}Fe_{30}$/GaAs structures possess good properties, with a high efficiency of spin transport observed in them [99]. Annealing of the Fe/GaAs structures at temperatures up to 300 °C results in a significant change of the physical properties of the Fe layers as well as in the spin dynamics.

## *5.4.2 Interlayer Exchange Coupling of the Magnetic Layers Based on GaMnAs*

In magnetic semiconductor structures, the phenomenon of the interlayer exchange coupling (IEC) has been found. To explain this phenomenon, the interaction mechanism suggested earlier for the description of DMS systems is commonly used [101]. According to the data of numerical calculation of a typical three-layer DMS system, between the magnetic layers, there must be a ferromagnetic or antiferromagnetic coupling, whose level depends on the thickness of the interlayer and the measurement temperature [102]. The first-principles calculation has shown that in the three-layer structures of GaMnAs/GaAs/GaMnAs with a spacer from GaAs doped with beryllium, the antiferromagnetic effect of IEC can be achieved due to a high level of doping with Be of the interfaces between GaMnAs and the spacer [103]. Switching from the ferromagnetic into antiferromagnetic state is possible. According to the data of calculation of the magnetic properties, one can produce antiferromagnetic coupling due to the orientation in a certain plane [104]. It should be noted that switching from the antiferromagnetic to the ferromagnetic state is possible under hole injection, which is important for semiconductor spintronics.

The calculation results have been confirmed in the experimental works [105–107]. In the three-layer structures of GaMnAs/GaAs/GaMnAs with the Mn content up to 3–5% and a spacer from undoped GaAs or that doped with Be with a thickness of up to 25 monolayers, a stable effect of the antiferromagnetic interlayer exchange coupling was observed. In the case of doping of GaAs with beryllium, there occurred a strong interaction, whereas in the undoped layers, a weak interaction. In the multilayer structures of GaMnAs/GaAs/Be with the nonmagnetic spacer GaAs/Be, the effect of the giant magnetoresistance was observed with high resistance in the weak fields and low resistance in the strong fields [106]. In these samples, one could also observe the negative coercive force and antiferromagnetic splitting. It was noted that in semiconductor multilayer structures, the antiferromagnetic interaction was stronger than in metal systems.

In the review [52], the authors have systematized the results on the interlayer exchange coupling in three-layer structures based on GaMnAs. The importance of the IEC for practical application is underlined. In a number of works [108–110], the heterostructures with the δ-layer doped with manganese have been studied. In the heterostructure of GaAs/AlGaAs with the δ-layer, the magnetic ordering was implemented due to the interaction between the magnetic Mn impurity and the two-dimensional hole gas [108]. In the structures of GaAs/InGaAs/GaAs with the δ-layer doped with Mn, ferromagnetic ordering was achieved at a small distance between the δ-layer and the quantum well (3–4 nm), which ensured an interaction between the Mn ions and the electrons in the quantum well [110].

In Ref. [109], the authors have investigated the spin polarization of the holes in the multilayer quantum wells of GaMnAs/AlAs. In their opinion, the hole polarization in the wells is determined by the properties of the main material rather than by the quantum environment.

The materials and structures in which magnetism is controlled by an electric field are called microferroics [42, 46, 111]. The authors in [47, 111] present a review on the properties of the heterostructures based on microferroics. Ferromagnetism in the GaMnAs DMS channel of the above structures is controlled by the nonvolatile field effect of spontaneous polarization. The magnetoelectric interaction in the above systems has the interface nature. In [47] the authors discuss the prospects for the use of the ferroelectric/DMS heterostructures to produce field transistors. The use of the ferroelectric polymer instead of the conventional ferroelectric oxide as a gate seems rather promising.

## 5.5 Application of Ferromagnetic Semiconductors

One of the trends of the development of modern semiconductor electronics is designing devices based on spin-dependent phenomena. It is assumed that on the basis of ferromagnetic semiconductors, it is possible to develop magnetic field sensors, reading heads for computers, galvanic insulators, magnetoresistive random access memory, and quantum computers [112–114]. According to the authors [115], by injecting spin-polarized charge carriers into a semiconductor, it is possible to implement qubits (quantum bits), i.e., operations necessary for quantum computers.

This is a significant step in the development of quantum computers based on the degree of freedom of the spin in solid-state devices. Diluted magnetic semiconductors can be used as spin injectors into a semiconductor. The magnetoresistive random access memory will allow decreasing energy consumption in electronics by an order of magnitude [112].

At the present time, there are some examples of practical application of ferromagnetic semiconductors and, first of all, gallium arsenide.

For fiber-optic communication, modern telecommunication systems require the magneto-optical materials with a large Faraday effect at a wavelength of 0.98 μm, which is determined by the applied fibers and compatibility with semiconductor devices. In their work [116], the authors have revealed a giant magneto-optical effect of the ferromagnetic nanoclusters of MnAs built in the GaAs matrix and shown the possibility of using it as a magneto-optical material. In this material, the Faraday rotation angle at a wavelength of 0.98 μm reaches $\approx 0.2°/\mu m$ at 2 kOe. The potential of the above material as the Faraday rotator is 16 times larger than that of the commercial CdMnHgTe developed as an optical isolator. Another merit of the given material is that its optical isolation is no less than 30 dB. However, its drawback is high internal losses.

The phenomenon of photo-enhanced magnetization in GaAs/Fe composite structures found by the authors of [72] was used to produce a light-driven microactuator [117].

The microactuator consists of a GaAs/Fe chip stuck to a Si cantilever. The chip is a multilayer structure of GaAs/(GaAs)m(Fe)n. In a magnetic field of 1.7 T with a chip irradiated by a laser at a wavelength of 650 nm and a power of 713 μW at room temperature, the cantilever deviates at a distance of 1 μm. Modern actuators are usually controlled by a change of the current through the electromagnet winding. The contactless control of actuators opens new possibilities for the development of wireless microelectronic mechanical systems. However, for practical application, it is necessary to increase the amplitude of photo-enhanced magnetization.

Field transistors (spin-FED) and light-emitting diodes (spin-LED) have been produced on the basis of the GaAs DMS [112, 118, 119]. In the spin-FED, the ferromagnetics connected by a narrow semiconductor channel serve as a source and a drain. A spin-polarized current flows from the source to the drain. If the source and the drain are magnetized in the same direction, an electric current flows between them, which corresponds to the true values of the Boolean domain. If they are magnetized in the opposite directions, the current decreases, which corresponds to false values. It allows implementation of a new efficient approach in designing logical devices.

Heterostructures with quantum wells and quantum dots on the basis of ferromagnetic AIIIBV semiconductors doped with manganese are used to manufacture light-emitting diodes [120, 121]. In the above structures, one can observe a rather high degree of circular polarization of luminescence indicating spin polarization of the nonequilibrium carriers with the QW, which can be used to transmit information with the help of spin codes [122].

In a number of works [123–126], light-emitting diodes based on the heterostructures with δ-layers doped with Mn have been investigated. It is shown that introduction of the Mn-doped δ-layers into the heterostructure of GaAs/InGaAs/GaAs with the undoped quantum well near the barrier of InGaAs/GaAs allows implementation of the circular polarization of the low-temperature electroluminescence of the diodes based on the above heterostructure. The authors have also studied the possibility of changing the degree of circular polarization of electroluminescence by varying the main design parameters (the spacer layer thickness, i.e., the distance between the δ-layer and the quantum well, the atom concentration in the δ-layer, the introduction of an additional acceptor δ-layer).

It has been established that the most efficient means of controlling the degree of circular polarization of electroluminescence is changing of the spacer layer thickness. It is also shown that introduction of a δ-layer results in an increase in the temperature stability and a decrease in the temperature quenching of luminescence in the structures under study as compared to the control samples without a δ-layer.

The temperature dependences of the photoluminescence of the heterostructures of InGaAs/GaAs with a quantum wall and the adjacent acceptor δ-layer doped with Mn have been studied in [119]. The structures of GaAsSb/GaAs containing a quantum well and δ-layers of manganese and carbon in the closing layer of GaAs have been grown and investigated [125]. At temperatures below 20–25 K, they possess the ferromagnetic properties. In Ref. [126], the authors have studied the influence of the method of growing the GaAs covering layer and building the Mn-doped δ-layer in the structure of InGaAs/GaAs with quantum points and wells on their crystalline and optoelectron characteristics. The quantitative chemical analysis of the structures of the spin light-emitting diodes with the spin-injecting layer of GaMnAs and the quantum well of InGaAs has been performed in [127] by X-ray photoelectron spectroscopy together with ion profiling. The method of phase separation and control of the correct determination of the component content has been improved. On the basis of the heterostructures of GaMnSb/InGaAs/GaAs, light-emitting diodes with a ferromagnetic injecting layer have been produced [128, 129]. Light-emitting diodes emit a circularly polarized light and possess efficient electroluminescence.

In Ref. [130], the authors have demonstrated the use of Fe/GaAs structures to produce a spin detector based on the magnetic Schottky barriers of Pd/Fe/GaAs. They have studied the transport of spin-polarized electrons in the above structures. An electron spin detector based on the Pd/Fe/GaAs structures can be used as a spin injector in solid-state spintronics.

To sum it up, this chapter is a review of the literature dealing with the production and properties of ferromagnetic gallium arsenide and the possibility of using it in spintronics. A study of ferromagnetic GaAs has been under way for two decades, and this material and the structures on its basis remain of great interest nowadays. It should be noted that along with some original works, there is a number of review articles [1, 2, 44, 45, 54, 112, 131–137] in which some or other properties of ferromagnetic GaAs are considered.

Gallium arsenide acquires the ferromagnetic properties when doped with the magnetic impurities of manganese or iron. Ferromagnetism in the AIIIBV compounds doped with manganese attracts great attention both from the physical and practical points of view. A number of the achievements presented in the above works indicate the key role of the (Ga,Mn)As DMS in integrating semiconductor and magnetic materials.

The nature of ferromagnetism is explained by the Zener exchange p–d model, which allows interpretation of the experimental results, including the Curie temperature, the anisotropy of ferromagnetism, and other phenomena. The influence of the antiferromagnetic interaction and self-compensation limit the Curie temperature for GaMnAs to about 200 K. The limited solubility of Mn in GaAs results in the formation of an ensemble of magnetic nanocrystals in the semiconductor matrix. The above heterogeneous systems have Tc higher than room temperature, and so they may become the basis for application in spintronics. Recently, there has been a growing interest in the hybrid structures combining the DMS with other ferromagnetic, antiferromagnetic, as well as nanostructured systems.

Ferromagnetism induced by charge carriers in such semiconductors as GaMnAs gives rise to a variety of the phenomena resulting from the interaction between the magnetic and semiconductor properties ensuring a "bridge" between a semiconductor and a ferromagnetic.

The prospects have been shown for application of GaAs doped with transition metals in spintronics. A number of devices have been made, e.g., field transistors and light-emitting diodes, a light-driven microactuator, magneto-optical materials, and magnetic field sensors. In the future, on the basis of the DMS, it will be possible to produce magnetic disk read-and-write heads and elements for quantum computers.

## References

1. Zutic I, Fabian J, Das Sarma S (2004) Spintronics: fundamentals and applications. Rev Mod Phys 76:323
2. Zakharchenya BP, Korenev VL (2005) Integrating magnetism into semiconductor electronics. Phys-Usp 48(6):603–608
3. Fert A (2008) Nobel lecture: origin, development, and future of spintronics. Rev Mod Phys 80 (4):1517–1530
4. Grunberg PA (2008) Nobel lecture: from spin waves to giant magnetoresistance. Rev Mod Phys 80(4):1531–1540
5. Baibich MN, Bruto JM, Fert A, Nguyen van Dau F, Petroff F, Eitenne P, Creuzet G, Friederich A, Chazelas J (1988) Giant magnetoresistance of (001)Fe/(001)Cr magnetic superlattices. Phys Rev Lett 61:2472
6. Binasch G, Grunberg P, Saurenbach F, Zinn W (1989) Enhanced magnetoresistance in layered magnetic structures with antiferromagnetic interlayer exchange. Phys Rev B 39:4828
7. Moodera JS, Kinder LR, Wong TM, Meservey R (1995) Large magnetoresistance at room temperature in ferromagnetic thin film tunnel junctions. Phys Rev Lett 74:3273
8. Ohno H (1998) Making nonmagnetic semiconductors ferromagnetic. Science 281:951
9. Kasuya T, Yanase A (1968) Anomalous transport phenomena in Eu-chalcogenide alloys. Rev Mod Phys 40:684
10. Furdyna J (1988) Diluted magnetic semiconductors. J Appl Phys 64:R29–R64

11. Ohno H, Munekata H, von Molnar S, Chang LL (1991) New III–V diluted magnetic semiconductors (invited). J Appl Phys 69:6103
12. Ohno H, Shen A, Matsukura F, Oiwa A, Endo A, Katsumoto S, Iye H (1996) (Ga, Mn)As: a new diluted magnetic semiconductor based on GaAs. Appl Phys Lett 69:363
13. Jungwirth T, Sinova J, Macek J, Kucera J, MacDonald AH (2006) Theory of ferromagnetic (III, Mn) V semiconductors. Rev Mod Phys 78:809
14. Dreher L, Bihler C, Peiner E, Waag A, Schoch W, Limmer W, Goennenwein STB, Brandt MS (2013) Angle-dependent spin-wave resonance spectroscopy of (Ga, Mn) As films. Phys Rev B 87:224422
15. Sliwa C, Dietl T (2014) Orbital magnetization in dilute ferromagnetic semiconductors. Phys Rev B 90:045202
16. He Z-X, Zheng H-Z, Xue-Jiao H, Hai-Long W, Zhao J-H (2014) What has been measured by reflection magnetic circular dichroism in $Ga_{1-x}Mn_xAs/GaAs$ structures. Chin Phys B 23 (7):077801
17. Kim K-M, Jho Y-S, Kim K-S (2015) Dilute magnetic topological semiconductors. Phys Rev B 91:115125
18. Szwacki NG, Majewski JA, Dietl T (2015) (Ga,Mn)As under pressure: a first-principles investigation. Phys Rev B 91:184409
19. Hayashi T, Katsumoto S, Hashimoto Y (2000) Anisotropy and Barkhausen jumps in diluted magnetic semiconductor (Ga,Mn)As. Physica B 284:1175
20. Yoona IT, Kanga TW, Kimb KH, Kimb DJ (2004) A Hall coefficient investigation of ferromagnetic $Ga_{1-x}Mn_xAs$ layers on (100) GaAs substrates. Solid State Comm 130:627
21. Hayashi T, Hashimoto Y, Katsumoto S, Iye Y (2001) Effect of low-temperature annealing on transport and magnetism of diluted magnetic semiconductor (Ga, Mn) As. Appl Phys Lett 78:1691
22. Hayashi T, Hashimoto Y, Yoshida S, Katsumoto S, Iye Y (2001) Control of material parameters and metal–insulator transition in (Ga, Mn) As. Phys E 10:130
23. Chiba D, Yamanouchi M, Matsukura F, Ohno H (2003) Electrical manipulation of magnetization reversal in a ferromagnetic semiconductor. Science 301:943
24. Novak V, Olejník K, Wunderlich J, Cukr M, Výborný K, Rushforth AW, Edmonds KW, Campion RP, Gallagher BL, Sinova J, Jungwirth T (2008) Curie point singularity in the temperature derivative of resistivity in (Ga, Mn) As. Phys Rev Lett 101:077201
25. Olejnik K, Owen MHS, Novak V, Mašek J, Irvine AC, Wunderlich J, Jungwirth T (2008) Enhanced annealing, high Curie temperature, and low-voltage gating in (Ga, Mn) As: a surface oxide control study. Phys Rev B 78:054403
26. Liu J (2014) Successful Mn ions spin polarization in magnetic semiconductor at room temperature in a $Co_2FeAl$/(Ga,Mn)As bilayer. Natl Sci Rev 1(1):3–4
27. Chen L, Yang X, Yang FH, Zhao JH, Misuraca J, Xiong P, Molnar SV (2011) Enhancing the Curie temperature of ferromagnetic semiconductor (Ga, Mn) As to 200 K via nanostructure engineering. Nano Lett 11:2584
28. De Biasia E, Pudenzia MAA, Beharc M, Carvasand F, Knobel M (2008) Study of the magnetic properties on Mn and As co-implanted GaAs. J Magn Magn Mater 320:404
29. Yokoyama M, Ogawa T, Nazmul AM, Tanaka M (2006) Large magnetoresistance (> 600%) of a GaAs: MnAs granular thin film at room temperature. J Appl Phys 99:08D502
30. Danilov YA, Kruglov AV, Pitirimova EA, Drozdov AV, Murel AV, Bekhar MA, Pudenzi MA (2004) Structure and properties of GaAs <Mn> layers formed by ion implantation. Izvestiya Akademii Nauk Ser Fizicheskaya [in Russian] 68(1):65–68
31. Zvonkov BN, Vikhrova OB, Danilov YA, Drozdov YN, Kudrin AB, Sapozhnikov MB (2010) Effect of compressive and tensile stresses in GaMnAs layers on their magnetic properties. Phys Solid State 52(11):2267–2270
32. Pereira ALJ, Dias da Silva JH (2008) Disorder effects produced by the Mn and H incorporations on the optical absorption edge of $Ga_{1-x}Mn_xAs$:H nanocrystalline films. J Non-Cryst Solids 354:5372

33. Angelico JC, Pereira ALJ, de Arruda LB, Dias da Silva JH (2015) Electrical transport mechanisms and structure of hydrogenated and non-hydrogenated nanocrystalline $Ga_{1-x}Mn_xAs$ films. J Alloy Comp 630:78
34. Bürger D, Zhou S, Pandey M, Viswanadham CS, Grenzer J, Roshchupkina O, Anwand W, Reuther H, Gottschalch V, Helm M, Schmidt H (2010) Application of pulsed laser annealing to ferromagnetic GaMnAs. Phys Rev B 81:115202
35. Piano S, Grein R, Mellor CJ, Vyborn K, Campion R, Wang M, Eschrig M, Gallagher BL (2011) Spin polarization of (Ga, Mn) As measured by Andreev spectroscopy: the role of spin-active scattering. Phys Rev B 83:081305(R)
36. Wadley P, Freeman AA, Edmonds KW, van der Laan G, Chauhan JS, Campion RP, Rushforth AW, Gallagher BL, Foxon CT, Wilhelm F, Smekhova AG, Rogalev A (2010) Element-resolved orbital polarization in (III, Mn) As ferromagnetic semiconductors from K-edge x-ray magnetic circular dichroism. Phys Rev B 81:235208
37. Uchitomi N, Sato S, Jinbo Y (2003) Growth and annealing effect of ferromagnetic (Ga, Mn) As on Si (100) substrates. Appl Surf Sci 216:607
38. Hamida AB, Sievers S, Pierz K, Schumacher HW (2013) Broadband ferromagnetic resonance characterization of GaMnAs thin films. J Appl Phys 114:123704
39. Hamida AB, Sievers S, Bergmann F, Racu A-M, Bieler M, Pierz K, Schumacher HW (2014) Magnetotransport and magnetization dynamics of GaMnAs thin films and magnetic tunnel junctions. Phys Status Solidi B 251(9):1652–1662
40. Jungwirth T, Wang KY, Masek J, Edmonds KW, Konig J, Sinova J, Polini M, Goncharuk NA, MacDonald AH, Sawicki M, Rushforth AW, Campion RP, Zhao LX, Foxon CT, Gallagher BL (2005) Prospects for high temperature ferromagnetism in (Ga, Mn) As semiconductors. Phys Rev B 72:165204
41. Dietl T, Ohno H, Matsukura F, Cibert J, Ferrard D (2000) Zener model description of ferromagnetism in Zinc-Blende magnetic semiconductors. Science 287:1019
42. Chiba D, Sawicki M, Nishitani Y, Nakatani Y, Matsukura F, Ohno H (2008. Sep. 25) Magnetization vector manipulation by electric fields. Nature 455(7212):515–518
43. Nishitani Y, Chiba D, Endo M, Sawicki M, Matsukura F, Dietl T, Ohno H (2010) Curie temperature versus hole concentration in field-effect structures of $Ga_{1-x}Mn_xAs$. Phys Rev B 81:045208
44. Ohno H (2013) Bridging semiconductor and magnetism. J Appl Phys 113:136509
45. Dietl T, Ohno H (2014) Dilute ferromagnetic semiconductors: physics and spintronic structures. Rev Mod Phys 86:187
46. Stolichnov I, Riester SWE, Mikheev E, Setter N, Rushforth AW, Edmonds KW, Campion RP, Foxon CT, Gallagher BL, Jungwirth T, Trodahl HJ (2011) Enhanced Curie temperature and nonvolatile switching of ferromagnetism in ultrathin (Ga, Mn) As channels. Phys Rev B 83:115203
47. Stolichnov I, Riester SWE, Mikheev E, Setter N, Rushforth AW, Edmonds KW, Campion RP, Foxon CT, Gallagher BL, Jungwirth T, Trodahl HJ (2011) Ferroelectric polymer gates for non-volatile field effect control of ferromagnetism in (Ga, Mn) As layers. Nanotechnology 22:254004
48. Riester SWE, Stolichnov I, Trodahl HJ, Setter N, Rushforth AW, Edmonds KW, Campion RP, Foxon CT, Gallagher BL, Jungwirth T (2009) Toward a low-voltage multiferroic transistor: magnetic GaMnAs under ferroelectric control. Appl Phys Lett 94:063504
49. Rader O, Fauth K, Gould C, Ruster C, Schott GM, Schmidt G, Brunner K, Molenkamp LW, Schutz G, Kronast F, Durr HA, Eberhardt W, Gudat W (2005) Identification of extrinsic Mn contributions in $Ga_{1-x}Mn_xAs$ by field-dependent magnetic circular X-ray dichroism. J Electron Spectrosc Relat Phenom 144:789
50. Hoffmann F, Woltersdorf G, Wegscheider W, Einwanger A, Weiss D, Back CH (2009) Mapping the magnetic anisotropy in (Ga, Mn) As nanostructures. Phys Rev B 80:054417
51. Winter A, Pascher H, Krenn H, Liu X, Furdyna JK (2010) Interpretation of hysteresis loops of GaMnAs in the framework of the Stoner–Wohlfarth model. J Appl Phys 108:043921

52. Chung S, Kim HC, Lee S (2009) The effect of carrier density on magnetic anisotropy of the ferromagnetic semiconductor (Ga, Mn) As. Solid State Commun 149:1739
53. Bihler C, Schoch W, Limmer W, Goennenwein STB, Brandt MS (2009) Spin-wave resonances and surface spin pinning in $Ga_{1-x}Mn_xAs$ thin films. Phys Rev B 79:045205
54. Lee S, Chung J-H, Liu X, Furdyna JK, Kirby BJ (2009) Ferromagnetic semiconductor GaMnAs. Mater Today 12:14
55. Woltersdorf G, Hoffmann F, Bauer HG, Back CH (2013) Magnetic homogeneity of the dynamic properties of (Ga,Mn) As films from the submicrometer to millimeter length scale. Phys Rev B 87:054422
56. Komamiya D, Okabayashi J, Yoshino J (2009) Mn deposition on GaAs (001)–c(4×4)α reconstructed surfaces: a scanning-tunneling-microscopy study. J Cryst Growth 311:2135
57. Sapega VF, Sablina NI, Panaiotti IE, Averkiev NS, Ploog KH (2009) Hole spin polarization in the exchange field of the dilute magnetic (Ga, Mn) As semiconductor studied by means of polarized hot-electron photoluminescence spectroscopy. Phys Rev B 80:041202(R)
58. Parchinskiy PB, Bobylev AY, Vlasov SI, Yu FC, Kim DJ (2007) Study of the photoluminescence spectra of GaMnAs produced by low-temperature molecular beam epitaxy. Semiconductors 41(10):1145–1149
59. Yastrubchak O, Zuk J, Krzyzanowska H, Domagala JZ, Andrearczyk T, Sadowski J, Wosinski T (2011) Photoreflectance study of the fundamental optical properties of (Ga, Mn) As epitaxial films. Phys Rev B 83:245201
60. Haisty RW (1965) On temperature depends of the iron acceptor level in GaAs. Appl Phys Lett 7:208
61. Omeljanovskii EM, Fistul VI (1983). Impurities of transition metals in semiconductors [in Russian]. M: Metallurgy: 192.
62. Harris IR, Smith NA, Cockayne B, MacEwan WR (1987) Phase identification in Fe-doped GaAs single crystals. J Cryst Growth 82:450
63. Isaev-Ivanov VV, Kolchanova NI, Masterov VF, Nasledov DN, Talalakin GN (1973) Magnetic properties of iron-doped gallium arsenide. Sov Phys Semicond [in Russian] 7 (2):299–300
64. Masterov VF, Samorukov BE (1978) Deep centers in III-V compounds (review). Sov Phys Semicond [in Russian] 12(4):363–381
65. Popov BP, Sobolevskii VK, Apushkinskii EG, Savel'ev BP (2005) Magnetic ordering effects in heavily doped GaAs:Fe crystals. Semiconductors 39(5):493–498
66. Chernov NA, Vilisova MD, Bakin NN, Asanov OM (1983) Determination of the position of the Π-ν-junction in epitaxial structures of gallium arsenide doped with iron. Sov Phys 26 (11):1043–1046
67. Vilisova MD, Ikonnikova GM, Moskovkin VA, Ruzaikin MP, Saprykin AI, Yabzanov VB, Yakubenya MP (1986) State of an Iron Impurity in the Gaseous and Solid Phases During Epitaxial Growth of Gallium Arsenide in the Ga-AsCl//3-H//2 System. Neorganiceskie materialy [in Russian] 22:363–366
68. Lavrent'eva LG, Vilisova MD (1989) Formation of centers with deep levels in gaseous phase epitaxy of gallium arsenide. Sov Phys J 29(5):339–347
69. Messmer R, Söderström E, Hult D, Marcinkevicius P, Lourdudoss S, Look S (2000) Properties of semi-insulating GaAs:Fe grown by hydride vapor phase epitaxy. J Electrochem Soc 147:3109
70. Hirose S, Yamaura M, Haneda S, Hara K, Munekata H (2000) GaFeAs: a diluted magnetic semiconductor grown by molecular beam epitaxy. Thin Solid Films 371:272
71. Haneda S, Yamaura M, Takatani Y, Hara K, Harigae S, Munekata H (2000) Preparation and characterization of Fe-based III–V diluted magnetic semiconductor (Ga, Fe)As. Jpn J Appl Phys Part 2 39:L9
72. Haneda S, Munekata H, Takatani Y, Koshihara S (2000) Fe-based magnetic-semiconductor hybrid structures for photocarrier-induced magnetism. J Appl Phys 87:6445

73. Soo YL, Kioseoglou G, Huang S, Kim S, Kao YH, Takatani Y, Haneda S, Munekata H (2001) Local structure around Fe in the diluted magnetic semiconductors $Ga_{1-x}Fe_xAs$ studied by X-ray absorption fine structure. Phys Rev B 63:195209
74. Park YJ, Oh HT, Park CJ, Cho HY, Shon Y, Kim EK, Moriya R, Munekata H (2002) Characteristics of molecular beam epitaxy-grown GaFeAs. Curr Appl Phys 2:379
75. Députier S, Guérin R, Lépine B, Guivarc'h A, Jézéquel G (1997) The ternary compound $Fe_3Ga_{2-x}As_x$: a promising candidate for epitaxial and thermodynamically stable contacts on GaAs. J Alloys Compd 262–263:416
76. Raghavan V (2004) As–Fe–Ga (Arsenic–Iron–Gallium). J Phase Equilib 25(1):77–78
77. Haneda S, Koshihara S, Munekata H (2001) Formation of FeAs and Fe crystallites in GaAs–Fe composite structures and their roles in light-enhanced magnetization. Phys E 10:437
78. Koshihara S, Iowa A, Hirasawa M, Katsumoto S, Iye Y, Urano C, Takagi H, Munekata H (1997) Ferromagnetic order induced by photogenerated carriers in magnetic III–V semiconductor Heterostructures of (In,Mn)As/GaSb. Phys Rev Lett 78:4617
79. Fumagalli P, Sommer G, Lippitz H, Haneda S, Munekata H (2001) Observation of reversed hysteresis loops and negative coercivity in granular GaAs–Fe hybrid structures. J Appl Phys 89:7016
80. Jamil ATMK, Noguchi H, Shiratori K, Kondo T, Munekata H (2005) Room-temperature photomagnetic effect of $Fe_3Ga_4$ grown on GaAs substrates. J Supercond 18(3):321–324
81. Jamil ATMK, Noguchi H (2011) Effect of thermal annealing of high temperature grows of $(GaAs_m(Fe)_n)_p$ composite films on GaAs (001) by molecular beam epitaxy. Indian J Phys 85:737
82. Changa JCP, Otsukab N, Harmon ES, Melloch MR, Woodall JM (1994) Precipitation in Fe- or Ni-implanted and annealed GaAs. Appl Phys Lett 65:2801
83. Taylor N, Sun K (2011) Fe–Ga–As precipitates and their magnetic domain structures in high-dose iron implanted GaAs. J Mater Sci 46:13
84. Khludkov SS, Prudaev IA, Novikov VA, Budnitskii DL, Lopatetskaya KS (2014) Electrical properties of GaAs doped with iron. Russ Phys J 56(12):1435–1438
85. Khludkov SS, Prudaev IA, Novikov VA, Tolbanov OP, Ivonin IV (2010) A study of the process of decomposition of supersaturated GaAs:Fe solid solution by scanning probe microscopy. Semiconductors 44(8):975–977
86. Prudaev IA, Khludkov SS, Gutakovskii AK, Novikov VA, Tolbanov OP, Ivonin IV (2012) Decomposition of a supersaturated solid solution of Fe in GaAs. Inorg Mater 48(2):93–95
87. Grunberg R, Schreiber Y, Pang M, Brodsky B, Sowers H (1986) Layered magnetic Structures: evidence for antiferromagnetic coupling of Fe layers across Cr interlayers. Phys Rev Lett 57:2442
88. Jamil ATMK, Noguchi H, Munekata H (2008) Growth of Fe–As crystalline films on GaAs (100) by molecular beam epitaxy. Thin Solid Films 516:3015
89. Prinz GA, Rado GT, Krebs JJ (1982) Magnetic properties of single-crystal {110} iron films grown on GaAs by molecular beam epitaxy (invited). J Appl Phys 53:2087
90. Krebs JJ, Jonker BT, Prinz GA (1987) Properties of Fe single crystal films grown on (100) GaAs by molecular beam epitaxy. J Appl Phys 61:2596
91. Tivakornsasithorn K, Liu X, Li X, Dobrowolska M, Furdyna JK (2014) Magnetic anisotropy in ultrathin Fe films on GaAs, ZnSe, and Ge (001) substrates. J Appl Phys 116:043915
92. Godde C, Noor S, Urban C, Köhler U (2008) Structural changes and alloying of annealed iron layers on GaAs(001) and GaAs(110). Surf Sci 602:3343
93. Chye Y, Huard V, White ME, Petroff PM (2002) Properties of a Fe/GaAs (001) hybrid structure grown by molecular-beam epitaxy. Appl Phys Lett 80:449
94. Lepin B, Lallaizon L, Schieffer P, Guivarc'h A, Jézéquel G, Rocher A, Abel F, Cohen C, Deputier S, Nguyen Van Dau F (2004) FeIIIGaAs (001): a stable and magnetic metal-semiconductor heterostructure. Thin Solid Films 446:6
95. Herfort J, Braun W, Trampert A, Schonherr H-P, Ploog KH (2004) Atomically engineered interfaces for spin injection: ultrathin epitaxial Fe films grown on As- and Ga-terminated GaAs (001) substrates. Appl Surf Sci 237:181

96. Torelli P, Sperl M, Ciancio R, Fujii J, Rinaldi C, Cantoni M, Bertacco R, Utz M, Bougeard D, Soda M, Carlino E, Rossi G, Back CH, Panaccione G (2012) Growth of ultrathin epitaxial Fe/MgO spin injector on (001) (Ga, Mn) As. Nanotechnology 23:465202
97. Bianco F, Bouchon P, Sousa M, Salis G, Alvaradoa SF (2008) Enhanced uniaxial magnetic anisotropy in Fe31Co69 thin films on GaAs. J Appl Phys 104:083901
98. Sacchi M, Marangolo M, Spezzani C, Breitwieser R, Popescu H, Dealaunay R, Salles BR, Eddrief M, Etgens VH (2010) Thermal switching of the magnetization in an iron film on a magnetically active template MnAs/GaAs (001). Phys Rev B 81:220401(R)
99. Salis S, Alvarado F, Fuhrer A (2011) Spin-injection spectra of CoFe/GaAs contacts: dependence on Fe concentration, interface and annealing conditions. Phys Rev B 84:041307(R)
100. Shaw JM, Falco CM (2007) Structure, spin dynamics, and magnetic properties of annealed nanoscale Fe layers on GaAs. J Appl Phys 101:033905
101. Dietl T, Haury A, d'Aubigne YM (1997) Free carrier-induced ferromagnetism in structures of diluted magnetic semiconductors. Phys Rev B 55:R3347(R)
102. Szałowski K, Balcerzak T (2009) Antiferromagnetic interlayer coupling in diluted magnetic thin films with RKKY interaction. Phys Rev B 79:214430
103. Luo M, Tang Z, Zhu ZQ, Chu JH (2011) Interlayer exchange coupling in GaN-based diluted magnetic semiconductor multilayers studied by first-principles calculations. J Appl Phys 109:123720
104. Huang GQ, Wang JX (2012) Magnetic behavior of Mn-doped GaN (1100) film from first-principles calculations. J Appl Phys 111:043907
105. Leiner J, Tivakornsasithorn K, Liu X, Furdyna JK, Dobrowolska M, Kirby BJ, Lee H, Yoo T, Lee S (2011) Antiferromagnetic exchange coupling between GaMnAs layers separated by a nonmagnetic GaAs: Be spacer. J Appl Phys 109:07C307
106. Chung S, Lee S, Chung J-H, Yoo T, Lee H, Kirby B, Liu X, Furdyna JK (2010) Giant magnetoresistance and long-range antiferromagnetic interlayer exchange coupling in (Ga,Mn) As/GaAs: Be multilayers. Phys Rev B 82:054420
107. Chung J-H, Song Y-S, Yoo T, Chung SJ, Lee S, Kirby BJ, Liu X, Furdyna JK (2011) Investigation of weak interlayer exchange coupling in GaMnAs/GaAs superlattices with insulating nonmagnetic spacers. J Appl Phys 110:013912
108. Nazmul AM, Sugahara S, Tanaka M (2003) Ferromagnetism and high Curie temperature in semiconductor heterostructures with Mn δ-doped GaAs and p-type selective doping. Phys Rev B 67:241308(R)
109. Sapega VF, Brandt O, Ramsteiner M, Ploog KH, Panaiotti IE, Averkiev NS (2007) Hole spin polarization in GaAs:Mn/AlAs multiple quantum wells. Phys Rev B 75:113310
110. Vaz CAF (2012) Electric field control of magnetism in multiferroic heterostructures. J Phys Condens Matter 24:333201
111. Morgunov RB, Dmitriev AI (2009) Spin dynamics in magnetic semiconductor nanostructures. Phys Solid State 51(10):1985–2002
112. Wolf SA, Awschalom DD, Buhrman RA, Daughton JM, von Molnár S, Roukes ML, Chtchelkanova AY, Treger DM (2001) Spintronics: a spin-based electronics vision for the future. Science 294:1488
113. Prinz ZA (1998) Magnetoelectronics. Science 282:1660
114. Di VDP (1995) Quantum computation. Science 270:255
115. Akinaga A, Miyanishi S, Tanaka K, Van Roy W, Onodera K (2000) Magneto-optical properties and the potential application of GaAs with magnetic MnAs nanoclusters. Appl Phys Lett 76:97
116. Shinshi T, Kato F, Shimokohbe A, Noguchi H, Munekata H (2003) Light-driven microcantilever actuator based on photoenhanced magnetization in a GaAs–Fe composite film. Appl Phys Lett 83:3425
117. Choi H-J, Seong H-K, Chang J, Lee K-I, Park Y-J, Kim J-J, Lee S-K, He R, Kuykendall T, Yang P (2005) Single-crystalline diluted magnetic semiconductor GaN:Mn nanowires. Adv Mater 17:1351

118. Dorokhin MV, Zaitsev SV, Brichkin AS, Vikhrova OV, Danilov YA, Zvonkov BN, Kulakovsky VD, Prokof'eva MM, Sholina AE (2010) Influence of delta⟨Mn⟩ doping parameters of the GaAs barrier on circularly polarized luminescence of GaAs/InGaAs heterostructures. Phys Solid State 52(11):2291–2296
119. Ohno Y, Young DK, Beschoten B, Matsukura F, Ohno H, Awschalom DD (1999) Electrical spin injection in a ferromagnetic semiconductor heterostructure. Nature 402:790
120. Taniyama T, Wada E, Itoh M, Yamaguchi M (2011) Electrical and optical spin injection in ferromagnet/semiconductor. NPG Asia Mater 3:65
121. Myers RC, Gossard AC, Awschalom DD (2004) Tunable spin polarization in III–V quantum wells with a ferromagnetic barrier. Phys Rev B 69:161305(R)
122. Dorokhin MB, Zaitsev SV, Brichkin AS, Vikhrova OB, Danilov YA, Zvonkov BN, Kulakovskii VD, Prokof'eva MM, Sholina AE (2010) Influence of parameters of the Mn Delta-doping GaAs barrier on circularly polarized luminescence of the heterostructures GaAs/InGaAs. Phys Solid State 52(11):2191–2296
123. Dorokhin MB, Danilov YA, Prokof'eva MM, Sholina AE (2010) Temperature stability of photoluminescence in heterostructures with InGaAs/GaAs quantum well and the acceptor Mn Delta-layer in the GaAs barrier. Tech Phys Lett 36(9):819–822
124. Zvonkov BN, Vikhrova OB, Danilov YA, Drozdov YN, Drozdov MN, Kalent'eva IL, Kudrin AV (2012) Fabrication and properties of GaAsSb/GaAs heterostructures doped with a magnetic impurity. Semiconductors 46(12):1493–1496
125. Pavlova ED, Gorshkov AP, Bobrov AI, Malekhonova NV, Zvonkov BN (2013) Study of heterostructures with a combined In(Ga)As/GaAs quantum dot/quantum well layer and a Mn δ layer. Semiconductors 47(12):1591–1594
126. Nikolichev DE, Boryakov AV, Zubkov SY, Kryukov RN, Dorokhin MV, Kudrin AV (2014) Chemical and phase composition of the spin-light emitting diodes GaMnAs/GaAs/InGaAs. Semiconductors 48(6):815–820
127. Dorokhin MV, Malysheva EI, Zdoroveischev AV, Danilov YA (2012) Spin-polarized light-emitting diodes based on heterostructures with a GaAs/InGaAs/GaAs quantum well and ferromagnetic GaMnSb injection layer. Tech Phys Lett 38(8):764–767
128. Dorokhin MV, Malysheva EI, Zdoroveischev AV, Danilov YA, Kudrin AV (2012) GaMnSb/InGaAs/GaAs heterostructure leds with a ferromagnetic injector layer. Semiconductors 46 (12):1518–1523
129. Kusrayev YG (2010) Spin phenomena in semiconductors: physics and applications. Physics-Uspekhi 53(7):725–738
130. Averkiev NS (2010) Spin relaxation anisotropy in two-dimensional semiconductors. Physics-Uspekhi 53(7):742–745
131. Tarasenko SA (2010) Spin photocurrents in semiconductors. Physics-Uspekhi 53(7):739–742
132. Jungwirth T, Wunderlich J, Novak V, Olejnik K, Gallagher BL, Campion RP, Edmonds KW, Rushforth AW, Ferguson AJ, Nemec P (2014) Spin-dependent phenomena and device concepts explored in (Ga,Mn)As. Rev Mod Phys 86:855
133. Tanaka M, Ohya S, Hail PN (2014) Recent progress in III–V based ferromagnetic semiconductors: band structure, Fermi level, and tunneling transport. Appl Phys Rev 1:011102
134. Khludkov SS (2008) Prospects for the use of gallium arsenide doped with transition metals as a material for spintronics [in Russian]. Russ Phys J 51(9/3):42–44
135. Khludkov SS (2010) Diluted magnetic semiconductors and devices based on spin-dependent phenomena [in Russian]. Russ Phys J 53(9/2):324–326
136. Khludkov SS, Prudaev IA, Tolbanov OP (2013) Gallium nitride as a material for spintronics. Russ Phys J 55(8):903–909
137. Khludkov SS, Prudaev IA, Tolbanov OP (2013) Prospects for the use of dilute magnetic semiconductors as a material for spintronics. In: Proceedings of XI Russian semiconductor physics symposium, Saint-Petersburg, 16–20 September, p 2

**Stanislav Khludkov** is a full professor and leading researcher of the Functional Electronics Laboratory in the Semiconductor Electronics Department of Tomsk State University, Russia. His research interests focus on the physical and electronic properties of GaAs and GaAs structures doped with deep-level impurities and the development of devices based on this material. He has authored more than 200 scientific publications and holds approximately 20 patents. Stanislav Khludkov can be reached by phone at +79627775632 or by email at knludkov@sibmail.com.

**Ilya Prudaev** is a senior researcher of Functional Electronics Laboratory in the Semiconductor Electronics Department of Tomsk State University, Russia. His research interests focus on the charge carrier transport in various semiconductor structures (GaAs, $Ga_2O_3$, $TiO_2$, AlGaInP, and AlGaInN heterostructures) as well as the development of electronic devices based on these materials. He has authored more than 50 scientific publications and holds 7 patents. Ilya Prudaev can be reached by phone at +79138868937 or by email at funcelab@gmail.com.

**Oleg Tolbanov** is a full professor and head of the Functional Electronics Laboratory in the Semiconductor Electronics Department of Tomsk State University, Russia. His research interests focus on the electronic properties of semiconductor structures doped with deep-level impurities and the development of devices. He is one of the creators of GaAs/Cr structures and matrix sensors. He has authored more than 300 scientific publications and holds approximately 50 patents. Oleg Tolbanov can be reached by phone at +79610984838 or by email at top@mail.tsu.ru.

# Chapter 6
# Nanosensors for Biomedical Applications: A Tutorial

**Holly Clingan, Alex Laidlaw, Pilarisetty Tarakeshwar, Micah Wimmer, Antonio García, and Vladimiro Mujica**

**Abstract** Nanoparticles coated with different kinds of molecules are currently designed and synthesized for several important applications, including catalysis, solar cells, and biomedical uses. A crucial molecular design variable is whether the nanoparticle exhibits plasmonic activity, e.g., the case of nanoparticles made of coinage metals, where no band gap is present, or if it rather behaves as a nano-semiconductor with a band gap, e.g., metal oxide nanoparticles. In this tutorial, we will discuss the literature for both plasmonic and non-plasmonic materials and our own recent theoretical and experimental work in two different showcases. First, we will present the example of using gold nanoparticles to monitor molecular sensing activity to follow changes in antibody/antigen binding through changes of the surface plasmon resonance (SPR) response. Second, we will discuss the case of surface-enhanced Raman resonance (SERS) in hybrid systems molecule-$TiO_2$ nanoparticles and clusters, where the important physical quantity is the Raman signal to monitor the formation of chemical bonds and interfacial electron transfer processes.

## 6.1 Introduction

The study of size-dependent properties of matter is a fascinating subject, because it challenges our physical and chemical intuition, which is mostly based on the behavior of the bulk material. What is found both experimentally is that properties of materials are strongly size-dependent. Thus, when particle size is in the nanoscale, optical, electric, chemical, and magnetic properties change as a function of the size of the particle. Such a dependence can be justified in terms of quantum models

H. Clingan (✉) · A. Laidlaw · P. Tarakeshwar · M. Wimmer · V. Mujica
School of Molecular Sciences, Arizona State University, Tempe, AZ, USA
e-mail: hclingan@asu.edu; tpilaris@asu.edu; micahwimmer@asu.edu; vmujica@asu.edu

A. García
Ira A. Fulton School of Engineering, Arizona State University, Tempe, AZ, USA
e-mail: tony.garcia@asu.edu



S. M. Goodnick et al. (eds.), *Semiconductor Nanotechnology*, Nanostructure Science and Technology, https://doi.org/10.1007/978-3-319-91896-9_6

because the distribution of energy levels and the geometry of a system are determined by the solution to Schrödinger equation, which in turn governs the electronic structure and the interatomic distances, key quantities in determining the properties of a material.

In this tutorial, we will mostly be interested in how the optical response of molecules is affected by their interactions with nanoparticles. In the first part, we will consider the case of plasmonic materials, e.g., gold and silver, which exhibit localized surface plasmons (LSP). The simplest way to describe plasmons is to consider them as excitations of an electron gas, which in this context is represented by the conduction electrons of the metal. LSP are non-propagating excitations of the conduction electrons of metallic nanostructures coupled to an external electromagnetic field. This response is resonant in nature as a function of the frequency of the field, which translates into an induced polarizability that exhibits a maximum, known as a localized plasmon resonance (PR). As a consequence, the radiation absorption intensity exhibits a maximum at the plasmon resonance, which is also correlated with the phenomenon of field amplification both inside and in the near-field zone outside the nanoparticle. This field enhancement is responsible for a variety of optical responses, e.g., fluorescence and Raman of molecules either chemically bonded or nonbonded to the metal nanoparticle. In the specific case of surface-enhanced Raman spectroscopy, the field enhancement factor of the molecular response can reach 12 orders of magnitude. This is a key element in the design of plasmonic sensors.

In the second part of this tutorial, we will consider the optical response of molecules chemically attached to the surface of a metal oxide, e.g., $TiO_2$. These materials are bulk semiconductors with a band gap that prevents the formation of plasmons, which requires the presence of delocalized conduction electrons, at least in their ground state. As a consequence, the field enhancement mechanism is absent in these nanoparticles, However, processes of interfacial charge transfer determine the appearance of chemically induced enhancement in the SERS response and a strong quenching of fluorescence in molecules attached to the nanoparticle, which are essential for the design of molecular senses using non-plasmonic materials.

## 6.2 Nanosensors

Very efficient sensors can be developed by using the properties of matter at the nanoscale. Several such sensors have been developed by taking advantage of changes in the melting point, fluorescence, electrical conductivity, magnetic permeability, and chemical reactivity, as well as the large surface to volume ratio of nanoparticles. These properties of nanomaterials can be used in generating a signal when analyte molecules interact with the nanoparticle, which acts as a functional unit in a sensing device.

The basic schematics for the design of a sensor are displayed in Fig. 6.1.

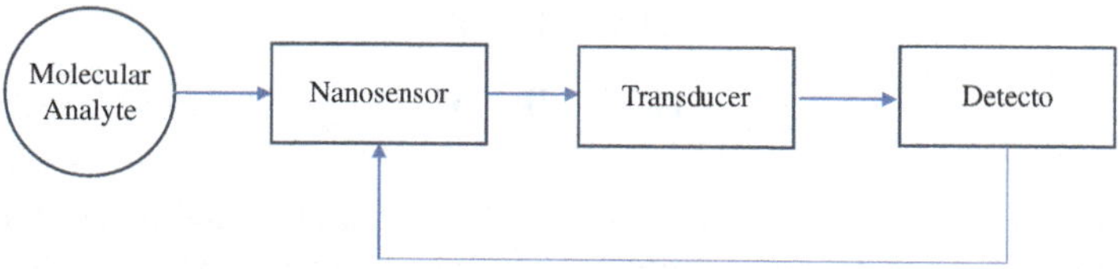


**Fig. 6.1** Schematic of a nanosensor

## 6.3 PR-based Sensors

Since field amplification occurs both inside and in the near-field zone outside the plasmonic nanoparticles, their use as sensors is possible even in the absence of chemical bond between the molecular analyte and the NP. But, before analyzing the changes in the plasmonic response of a NP due to the presence of a molecular analyte, which is our main objective here, it is important to have a basic description of the optical response of a plasmonic NP to an external electromagnetic field. In what follows we follow closely the treatment in S. A. Maier's book, *Plasmonics: Fundamentals and Applications*. Springer 2007.

### *6.3.1 Dielectric Function and Conductivity*

The first step to describe the optical response of a material is to establish a connection between the dielectric function and the conductivity. We start by recalling Maxwell's equation, which governs the electromagnetic response and field propagation. We write these fundamental equations in the following form:

$$\nabla.\mathbf{D} = \rho_{\text{ext}} \tag{6.1a}$$
$$\nabla.\mathbf{B} = 0 \tag{6.1b}$$
$$\nabla \times \mathbf{E} = -\frac{\partial \mathbf{B}}{\partial \mathbf{t}} \tag{6.1c}$$
$$\nabla \times \mathbf{H} = \mathbf{J}_{\text{ext}} + \frac{\partial \mathbf{D}}{\partial t} \tag{6.1d}$$

where **D**, **B**, **E**, and **H** are the dielectric displacement, the magnetic induction, the electric field, and the magnetic field. These equations link the four macroscopic fields, with the external charge, $\rho_{\text{ext}}$, and the current density $\mathbf{J}_{\text{ext}}$. This form of writing Maxwell's equations is convenient for the description of optical response, because the charge and the current density are divided as sums of internal and external contributions, that is, $\rho_{\text{tot}} = \rho_{\text{ext}} + \rho$, and $J_{\text{tot}} = J_{\text{ext}} + J$. In this description, the external components drive the system, while the internal set responds to the external stimuli.

The two auxiliary fields, **D** and **H**, are defined by

$$\mathbf{D} = \varepsilon_o \mathbf{E} + \mathbf{P} \tag{6.2a}$$

$$\mathbf{H} = \frac{1}{\mu_o} \mathbf{B} - \mathbf{M} \tag{6.2b}$$

where P is the polarization, M is the magnetization, and $\varepsilon_o$ and $\mu_o$ are the electric permittivity and magnetic permeability of vacuum, respectively.

For linear, isotropic, and nonmagnetic media, one can introduce the constitutive relations

$$\mathbf{D} = \varepsilon_o \varepsilon \mathbf{E} \tag{6.3a}$$

$$\mathbf{B} = \mu_o \mu \mathbf{B} \tag{6.3b}$$

In these equations $\varepsilon$ is the dielectric constant, and $\mu$ is the relative permeability. The linear relationship between **D** and **E** is equivalent to the equation used in describing the optical response of a material under the influence of an external electric field

$$\mathbf{P} = \varepsilon_o \chi \mathbf{E} \tag{6.4}$$

where $\chi$ is the dielectric susceptibility. Inserting (6.4) and (6.2a) into (6.3a) yields the important equation $\varepsilon = 1 + \chi$.

In the linear response regime, another important relationship between the internal current density and the electric field can be established, by introducing the conductivity $\sigma$.

$$\mathbf{J} = \sigma \mathbf{E} \tag{6.5}$$

To describe the optical response of metals, one needs to generalize Eqs. (6.3a) and (6.5) to take into account the non-locality in time and space. This leads to considering the general equations

$$\mathbf{D}(\mathbf{r}, t) = \varepsilon_o \int dt' dr' \varepsilon\left(\mathbf{r} - \mathbf{r}', t - t'\right) E\left(\mathbf{r}', t'\right) \tag{6.6a}$$

$$\mathbf{J}(\mathbf{r}, t) = \varepsilon_o \int dt' dr' \sigma\left(\mathbf{r} - \mathbf{r}', t - t'\right) E\left(\mathbf{r}', t'\right) \tag{6.6b}$$

taking the Fourier transform of these two fields

$$\mathbf{D}(\mathbf{k}, \omega) = \int dt dr \mathbf{e}^{i(\mathbf{k}.\mathbf{r} - \omega t)} \mathbf{D}(\mathbf{r}, t) \tag{6.7a}$$

$$\mathbf{J}(\mathbf{k}, \omega) = \int dt dr \mathbf{e}^{i(\mathbf{k}.\mathbf{r} - \omega t)} \mathbf{J}(\mathbf{r}, t) \tag{6.7b}$$

Using Eqs. (6.7a) and (6.7b) together with Eqs. (6.2a) and (6.2b), it is possible to derive a fundamental relationship between the dielectric function and the conductivity.

$$\varepsilon(\mathbf{k},\omega) = 1 + \frac{i\sigma(\mathbf{k},\omega)}{\varepsilon_0\omega} \tag{6.8}$$

## Drude Model for a Free Electron Gas

The simplest model for a metal consists of treating conduction electrons as a classical "gas" of free electrons with density $n$. The electrons in this plasma oscillate in response to the field, and their motion is damped due to collisions occurring with a characteristic collision frequency $\gamma = 1/\tau$, where $\tau$ is the relaxation time of the free electron gas.

The Newton equation of motion for an electron of the plasma under the influence of an external field $\mathbf{E}$ is then

$$m\ddot{\mathbf{x}} + m\gamma\dot{\mathbf{x}} = -e\mathbf{E} \tag{6.9}$$

The displaced electrons contribute to the macroscopic polarization

$$\mathbf{P} = -ne\mathbf{x} \tag{6.10}$$

Assuming a simple field of the form $\mathbf{E}(t) = \mathbf{E}_o e^{-i\omega t}$, and solving Eq. (6.9), it can be calculated to be

$$\mathbf{P} = -\frac{ne^2}{m(\omega^2 + i\gamma\omega)}\mathbf{E} \tag{6.11}$$

Inserting this equation into Eq. ((6.2a), we obtain

$$\mathbf{D} = \varepsilon_o\left(1 - \frac{\omega_p^2}{m(\omega^2 + i\gamma\omega)}\right)\mathbf{E} \tag{6.12}$$

where $\omega_p^2 = \frac{ne^2}{\varepsilon_o m}$ is the plasma frequency of the electron gas. The frequency-dependent complex dielectric function of the electron gas is then

$$\varepsilon(\omega) = 1 - \frac{\omega_p^2}{\omega^2 + i\gamma\omega} \tag{6.13}$$

## Mie Theory

A semiclassical description of the response of a metal NP to an electromagnetic field is given by Mie theory, which is an appropriate level of description of many applications where the details of the quantum description are not required. We use here the metallic nanoparticle boundary element method (MNPBEM) toolbox version of Mie model.

Quasistatic Approximation

The primary aim of the MNPBEM toolbox is solving Maxwell's equations using boundary conditions of the particles' surface in order to calculate the stimulated electromagnetic fields.

A quasistatic approximation can be implemented when the particles are much smaller than the wavelength of the external electromagnetic field.

The electrostatic potential is solved for while maintaining full frequency-dependent dielectric functions when considering boundary conditions by solving the Poisson-Laplace equation. This is an unbounded, homogenous point surface solution of the Poisson equation including electrostatic Green function:

$$\nabla^2 G(\mathbf{r},\mathbf{r}') = -4\pi\delta(\mathbf{r}-\mathbf{r}') \tag{6.14}$$

$$G(\mathbf{r}-\mathbf{r}') = \frac{1}{|\mathbf{r}-\mathbf{r}'|} \tag{6.15}$$

where G represents the Green function from the Coulomb equation and **r** represents the distance from the center of mass.

In this case we have an inhomogeneous dielectric environment, so the boundaries are split between the gold nanoparticle and the surrounding medium.

$$\phi(\mathbf{r}) = \phi_{ext}(\mathbf{r}) + \oint_{V_i} G(\mathbf{r},\mathbf{s})\rho(\mathbf{s})da \tag{6.16}$$

$\phi$ and $\phi_{ext}$ represent the electrostatic potential and the external electrostatic potential, respectively. $\rho(\mathbf{s})$ is the surface charge distribution located at the particle boundary, $\rho(\mathbf{s})$ is the surface charge distribution located at the boundary, $\partial Vi$. $\frac{\partial}{\partial n}$ denotes the derivative along the direction of the outer surface normal, and ε1 and ε2 are the dielectric functions of the particles and the medium that it sits, respectively.

$$\Lambda\rho(\mathbf{s}) + \oint \frac{\partial G(\mathbf{s},\mathbf{s}')}{\partial n}\rho(\mathbf{s})da' = -\frac{\partial\phi_{\text{ext}}(\mathbf{s})}{\partial n}, \Lambda = 2\pi\frac{\varepsilon_2+\varepsilon_1}{\varepsilon_2-\varepsilon_1} \tag{6.17}$$

This is used to elucidate the surface charge distribution $\rho$.

$\Lambda$ is governed by the dielectric functions in- and outside the particle boundaries. $\varepsilon_1$ and $\varepsilon_2$ represent the two dielectric functions, the first represents inside of the gold nanoparticle and the second the medium, water, in which the particle sits.

$$\Lambda\rho_i + \sum\nolimits_j \left(\frac{\partial G}{\partial n}\right)_{ij}\rho = -\left(\frac{\partial\phi_{\text{ext}}}{\partial n}\right)_i \tag{6.18}$$

ρ represents surface charges which can be elucidated by simple matrix inversion. The surface derivative $\left(\frac{\partial G}{\partial n}\right)_{ij}$ of the Green function connects surface element $i$ and $j$, and the surface derivative $\left(\frac{\partial G}{\partial n}\right)_{ij}$ of the external potential.

Full Maxwell's Equations

### Helmholtz Equation and Green Function

$$\left(\nabla^2 + k_i^2\right)G_i(\boldsymbol{r},\boldsymbol{r}') = -4\pi\delta(\boldsymbol{r}-\boldsymbol{r}'), G_i(\boldsymbol{r},\boldsymbol{r}') = \frac{e^{ik_i|\boldsymbol{r}-\boldsymbol{r}'|}}{|\boldsymbol{r}-\boldsymbol{r}'|} \tag{6.19}$$

$k_i = \sqrt{\varepsilon_i}k$ is the wavenumber in the medium.
$k = \frac{\omega}{c}$ is the wavenumber in a vacuum and c is the speed of light.
The magnetic permeability μ is set to one throughout.

### For an Inhomogeneous Dielectric Environment

$$\phi(\boldsymbol{r}) = \phi_{ext}(\boldsymbol{r}) + \oint_{V_i} G_i(\boldsymbol{r},\boldsymbol{s})\sigma_i(\boldsymbol{s})da \tag{6.20}$$

$$\boldsymbol{A}(\boldsymbol{r}) = \boldsymbol{A}_{ext}(\boldsymbol{r}) + \oint_{V_i} G_i(\boldsymbol{r},\boldsymbol{s})\boldsymbol{h}_i(\boldsymbol{s})da \tag{6.21}$$

This fulfills the Helmholtz equations everywhere except at the particle boundaries. σ*i* and *hi* are surface charge and current distributions, and φext and *A*ext are the scalar and vector potentials characterizing the external perturbation. The scalar and vector potentials are additionally related through the Lorentz gauge condition $\nabla \cdot \mathbf{A} = ik\varepsilon\phi$.

Dielectric Environment

This defines that the particle is closed by the following summation rule.

$$\oint_{\partial V_i} \frac{\partial G(s,s')}{\partial n} da' = 2\pi, \sum_j \left(\frac{\partial G}{\partial n}\right)_{ij} = 2\pi \tag{6.22}$$

An oscillating dipole ***d*** is assumed to be located in the vicinity of a metallic nanosphere, and the corresponding equations in Mie theory need to be adjusted to calculate the resulting electric field.

$$\boldsymbol{E}(\boldsymbol{r}) = k^2\boldsymbol{G}(\boldsymbol{r},\boldsymbol{r}';\omega).\boldsymbol{d} \tag{6.23}$$

$$k = \frac{\omega}{c} \tag{6.24}$$

The resulting physical description can be used to model changes in the plasmonic response with a metal NP due to both the presence of a dipole that represents a discrete structure mimicking a molecule and a solvent characterized by a dielectric constant.

### 6.3.2 A Showcase: Using PR to Diagnose Infectious Diseases

The design of new nanosensors for biomedical applications requires a concerted theory-experiment to optimize both the fabrication conditions and the interpretation of the measurements within a relatively simple conceptual framework [1–3]. As mentioned in the introduction, the general strategy for the design of nanosensors based on photonic materials such as the noble metals involves using the plasmonic resonance as basic optical signal, whose intensity, shift, and splitting characterize the response of the system [4–6]. Changes in the plasmonic response are due to combination of chemical and physical factors, e.g., molecule-nanoparticle charge transfer, or changes in the dielectric constant of the medium associated with the coating of the nanoparticle with the chemical species involved in the detection process [5, 7–11]. For biomedical applications, the coating agent is frequently one of the two members of the antigen-antibody pair involved in the immune response [3, 12, 13].

In this section, we describe the main results of our investigation about the interaction between bovine serum albumin (BSA) and gold colloids using ultraviolet (UV) and visible light absorption spectroscopy measurements to determine the surface coverage and binding activity [4–6, 13, 14]. The binding of BSA to the ubiquitous citrate-coated gold nanoparticles (AuNP) suggests an electrostatic interaction mechanism [11, 13, 15]. Surface coverage on the colloids is based on the concentration of BSA [10]. The measurements of the surface plasmon resonance (SPR) show that BSA and citrate-coated AuNP achieve stabilization and surface coverage at or above the isoelectric point (~4.7) and that the optical response of the system corresponds to a change in intensity only of the SPR [4–6, 8, 10]. The data supports a non-covalent and non-spontaneous binding mechanism of gold colloids and shows a maximum surface coverage that is dependent on concentration [10, 13].

Our hypothesis is that AuNP flip surface charges of the antibodies and bind non-covalently due to electrostatic and hydrophobic interactions [8, 10, 13, 16]. This causes the SPR intensity to decrease because of the monolayer coating disrupting the signal [13, 17, 18]. To achieve this non-covalent binding mechanism, the pH of the buffer used for conjugation to the gold nanoparticles must have a negative net charge [2, 8]. Usually this means that the pH is above the isoelectric point (pI) of the protein or antibody and for BSA this value is ~4.7, so the buffer must be above a pH of

5 [2, 8, 10, 11, 13, 15]. The ubiquitous citrate coating of the AuNPs in $dH_2O$ ensures that the net charge remains negative.

The results indicate that a simple theoretical framework, based on Mie theory, can explain the most important experimental trends. The key feature is that the optical response depends on choosing a dielectric constant, which in turn is determined by the conditions of electrostatic equilibrium [5, 8]. The MNPBEM toolbox is a flexible software that can simulate metallic nanoparticles, specifically gold nanoparticles using boundary element method [19]. The toolbox will set up homogeneous, isotropic dielectric functions with strict boundaries, which will then be used to calculate and solve Maxwell's equations and compare to Mie theory. By writing, manipulating, and altering the coded data within the toolkit, a very reasonable representation of the experimental system can be achieved.

## 6.4 Experimental Results

### *6.4.1 Methods*

#### Conjugation of AuNP and BSA

Three standard bovine serum albumin (BSA – 1 mg/ml stock solution) solutions were prepared and diluted with concentrations ranging from 1 mg/ml to 1 μg/ml. This was done by adding 10 mg BSA to 10 ml of $dH_2O$ to a 15 ml centrifuge tube. Similarly, a second and third dilution was done by adding 9 ml of $dH_2O$ to 15 ml centrifuge tubes and adding 1 ml of the 1 mg/ml BSA solution and 0.1 mg/ml BSA solution, respectively, for total volumes of 10 ml. Gold nanoparticles from Ted Pella, Inc. were aliquoted in 2 ml Eppendorf tubes with 1 ml of 20 nm AuNP containing $7 \times 10^{10}$ particles/ml and 60 nm AuNP containing $2.6 \times 10^{10}$ particles/ml in 0.1 M phosphate buffer (pH = 7). With 0.2 ml PCR tubes, 6 concentrations of BSA and the AuNP with phosphate buffer were prepared ranging from 0.024 mg/ml to 0.2 mg/ml. The tubes incubated at 37 ° C for 10 min. Each concentration was measured in triplicate using a UV-visible light spectrophotometer.

#### UV-Vis Spectroscopy and Light Scattering

The samples were measured using Ocean Optics USB4000 UV-Vis spectrometer between 400 and 800 nm using SpectraSuite Software for control and data acquisition. A quartz cuvette of 1 mm path length was used to acquire every sample. A deuterium-halogen light source was used to collect the UV-Vis spectra. The data collection was collected at 1 s integration time, and 5 spectra were averaged for each sample. The drift in spectral recordings was accounted for by normalizing the measurements in Plot2 scientific 2D plotting program. Any Mie theory calculations

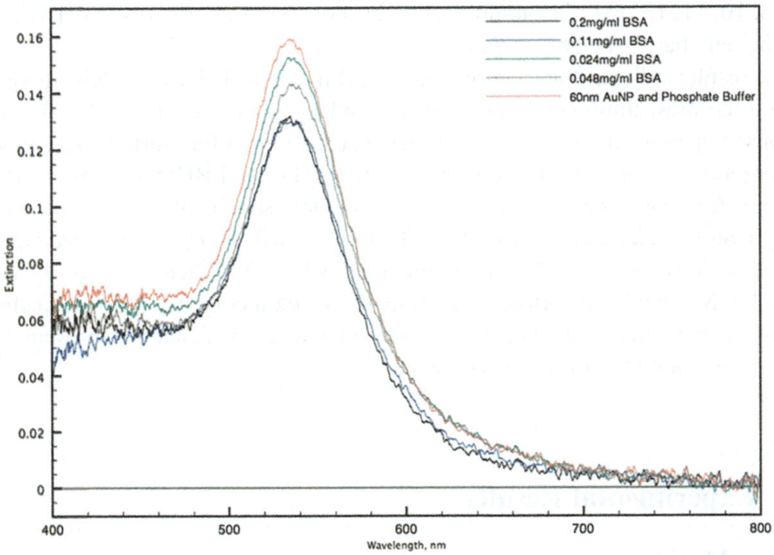


**Fig. 6.2** Absorbance spectroscopy for 60 nm AuNP with phosphate buffer in a water solvent and BSA concentrations ranging from 0.024 mg/ml to 0.2 mg/ml

for light extinction were made with the Mie theory calculator from the nanocomposix website (http://nanocomposix.com/pages/tools).

## Plots and Experimental Results

The experimental results followed the predicted hypothesis that asserted as concentration increases the SPR intensity decreases. Figure 6.2 shows the 60 nm AuNP with BSA concentrations ranging from 0.024 mg/ml to 0.2 mg/ml. Not shown here, the 20 nm AuNP with BSA also yielded very similar results along with the 50% by volume glycerol/water as the solvent.

Figure 6.3 is the first derivative of the absorbance spectroscopy plot from Fig. 6.2. This plot reduces noise in the spectrophotometry caused by scattering and characterizes the rate change of the absorbance with respect to the wavelength [6, 20]. Additionally, $\lambda_{max}$ of the absorbance band also passes through zero at the same wavelength [6]. This correlates to the minimum and maximum of the spectra being inflection points in that absorbance band [6].

To further analyze this correlation, $\lambda_{max}$ was taken against the concentration of BSA and 60 nm AuNP as shown in Fig. 6.4. The intensity of the SPR decreases as the concentration of BSA increases, and the intensity drop plateaus once the AuNP are coated. The application of these results, as previously mentioned, can be applied to biosensing technologies that take advantage of these properties for drug delivery, amplification of immunosensors, and rapid ELISAs [21–23].

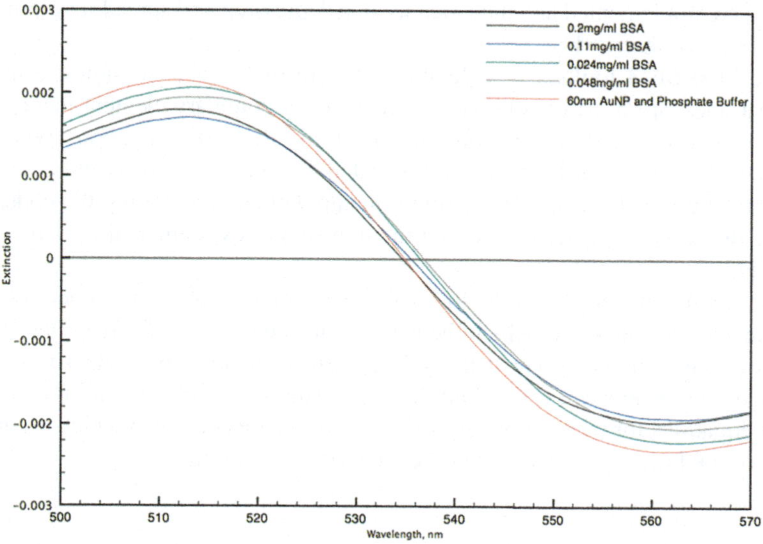


**Fig. 6.3** First derivative of the absorbance spectroscopy for 60 nm AuNP with phosphate buffer in a water solvent and BSA concentrations ranging from 0.024 mg/ml to 0.2 mg/ml

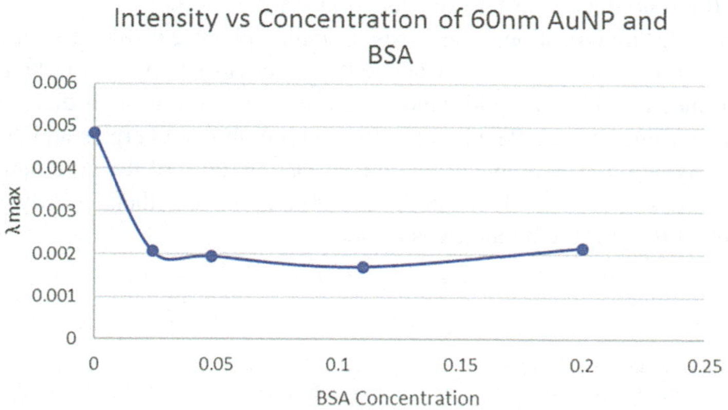


**Fig. 6.4** λmax (intensity) versus BSA concentration with 60 nm AuNP

## Computational Model

The computational component models the plasmonic response of gold nanoparticles of 60 nm and 20 nm diameter in a water solution and again in a water-glycerol solution. These are the control experiments to assess the plasmonic response of the gold colloids without the interactions of the BSA. The experimental results reveal that the peak plasmonic response wavelengths for the 60 nm and 20 nm gold nanoparticles are 534 nm and 523 nm, respectively. As increasing concentrations of BSA are implemented and then increased, a depression and spreading of this peak wavelength is observed.

The MNPBEM toolbox is a flexible software that is able to simulate metallic nanoparticles, specifically gold nanoparticles using boundary element method. The toolbox will set up homogeneous and isotropic dielectric functions with strict boundaries, which will then be used to calculate and solve Maxwell's equations and compare to MIE theory. By writing, manipulating, and altering the coded data within the toolkit, an accurate representation of the experimental system can be modelled.

The simulation consists of the following steps: first the dielectric functions of the particle and the solution in which the particle sits are defined. The particles, surface features, and boundaries are then outlined. Subsequently, the code specifies the nature of the plane wave excitation; this information is used to input into the BEM solver equations for which auxiliary surface charges are computed which produces a graph of total decay rate vs. plasmon excitation wavelength.

### *Description of Results and Techniques Used to Obtain Results*

The BEMMNP toolkit uses boundary element method which defines systems with strict borders and well-defined dielectric functions [19]. The system that is outlined is for a 60 nm gold nanoparticle embedded in water solution.

Figure 6.5, "initialization," describes a single sphere embedded in a homogeneous dielectric environment. The water dielectric environment is described using its refractive index, 1.33. The gold nanoparticle is expressed through dielectric functions whose values are tabulated on file for specific photon energies and performs a spline interpolation. The diameter and the surface vertices of the gold nanoparticle are then outlined by use of the polar coordinates pi and theta. Lastly the code specifies that the particle boundary is closed.

```
%% initialization
addpath( genpath( '/Users/AlexLaidlaw/Desktop/MNPBEM17' ) );
% options for BEM simulation
op = bemoptions( 'sim', 'ret', 'interp', 'curv' );

% table of dielectric functions
epstab = { epsconst(1.33^2), epstable( 'gold.dat' ) };
% diameter of sphere
diameter = 60;
% nanosphere with finer discretization at the top
phi=linspace(0,2*pi,42);
theta=linspace(0,pi,30);
p = trispheresegment( phi,theta,diameter,'triangles');
% initialize sphere
p = comparticle( epstab, { p }, [ 2, 1 ], 1, op );
```

**Fig. 6.5** Initialization code for resonance response of gold nanoparticle when embedded in water

```
%% dipole oscillator
enei = linspace( 300, 900, 100 );
% compoint
pt = compoint( p, [ 0, 0, 0.7 * diameter ] );
% dipole excitation
dip = dipole( pt, [ 1, 0, 0; 0, 0, 1 ], op );
% initialize total and radiative scattering rate
[ tot, rad ] = deal( zeros( numel( enei ), 2 ) );
```

**Fig. 6.6** Dipole oscillator code for resonance response of gold nanoparticle when embedded in water

```
%% BEM simulation
% set up BEM solver
bem = bemsolver( p, op );

multiWaitbar( 'BEM solver', 0, 'Color', 'g', 'CanCancel', 'on' );
% loop over wavelengths
for ien = 1 : length( enei )
  % surface charge
  sig = bem \ dip( p, enei( ien ) );
  % total and radiative decay rate
  [ tot( ien, : ), rad( ien, : ) ] = dip.decayrate( sig );

  multiWaitbar( 'BEM solver', ien / numel( enei ) );
end
% close waitbar
multiWaitbar( 'CloseAll' );
```

**Fig. 6.7** BEM simulation code for resonance response of gold nanoparticle when embedded in water

Figure 6.6 sets up an oscillating dipole along the z axis where the energies are defined by the wavelength of the dipole. We test from 400 nm to 900 nm as this wavelength range corresponds to the UV-visible spectrum of light (Fig. 6.7).

The BEM solvers use the particle boundaries and Maxwell's equation to solve for the surface charges and currents. This results in the total and radiative decay of the plasmonic response in the gold nanoparticles to be calculated. The final plots are then coded for but are omitted from this report (Fig. 6.8).

Finally, the MNPBEM toolkit allows the setup of an identical system in which it solves for the total radiative decay over energy, for comparison, by Mie theory [19]. Mie theory is a solution to Maxwell's equations which accurately describe the scattering of an electromagnetic plane wave by a sphere, for example, a gold nanoparticle.

Figure 6.9 represents the plasmonic response of a 60 nm gold nanoparticle embedded in water and is the output of the code. The maximum decay rate occurs at 530 nm and at a total decay rate of 68. The plot of the 20 nm gold nanoparticle has

```
%% comparison with Mie theory
mie = miesolver( epstab{ 2 }, epstab{ 1 }, diameter, op );
  total and radiative decay rate
[ tot0, rad0 ] = deal( zeros( numel( enei ), 2 ) );

% loop over energies
for ien = 1 : numel( enei )
  [ tot0( ien, : ), rad0( ien, : ) ] = ...
    mie.decayrate( enei( ien ), pt.pos( :, 3 ) );
end

if ~verLessThan( 'matlab', '8.4.0' ), set( gca, 'ColorOrderIndex', 1 ); end
plot( enei, tot0, '--' );
if ~verLessThan( 'matlab', '8.4.0' ), set( gca, 'ColorOrderIndex', 1 ); end
plot( enei, rad0, 'o--' );

legend( 'tot(x) @ BEM', 'tot(z) @ BEM', ...
        'rad(x) @ BEM', 'rad(z) @ BEM', 'tot(x) @ Mie' );
```

**Fig. 6.8** Mie theory comparison code for resonance response of gold nanoparticle when embedded in water

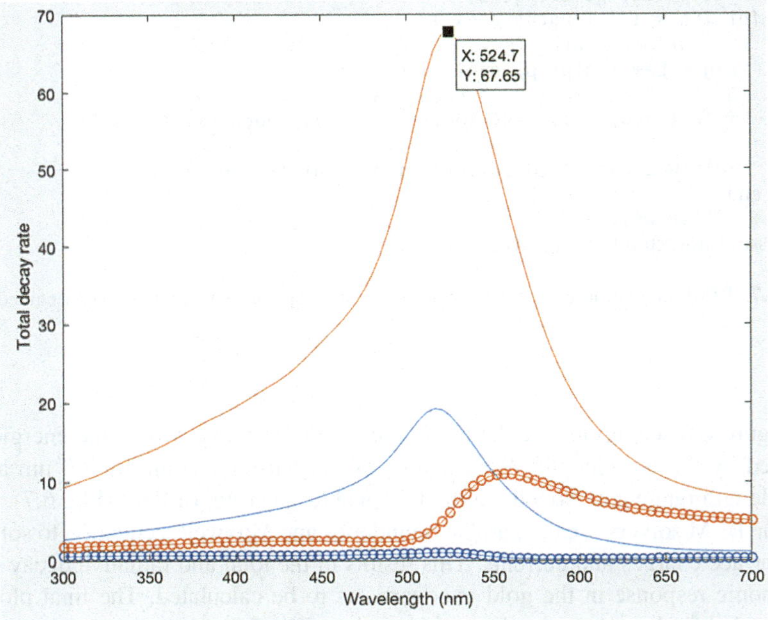


**Fig. 6.9** Total decay rate vs. wavelength resonance response of a 60 nm gold nanoparticle when embedded in water

a very similar aesthetic, with a much greater decay rate and marginally lower maximum wavelength peak. Current results also show that as the BSA antibody is introduced the maximum peak shifts to higher wavelengths (lower energies) and that the total decay rate reduces.

## 6.5 SERS-Based Sensors in Molecule-Metal Oxide Hybrids

Hybrid nanosystems formed by a molecule (M) and a nanoparticle (NP) that is either a bulk-phase metal (MNP) or a semiconductor (SCNP) are of fundamental importance in sensors, photonics, catalysis, and photovoltaics devices [24–35], just to mention a few areas of current interest. An important reference for our work is single-molecule SERS in M-MNP systems a technique that was developed a few years ago [36–39]. The fundamental physics of the enhancement is related to a Raman transition in the presence of a giant dipole field that arises as of plasmon resonance mediated increase of the local electromagnetic field and an increase in the polarizability. The combined effect can be very substantial, more than 6 orders of magnitude, and has made SM-SERS an extremely sensitive molecular analysis and sensing technique.

Hybrid M-SCNP are of great importance for the design of photovoltaic devices [32], e.g., DSSC (dye-sensitized solar cells) [40, 41], artificial photosynthesis [42–45], photocatalysts [46, 47], and biomolecular sensors [24–28]. A critical and largely unresolved issue in this field is the understanding of the charge transfer (CT) interfacial properties and the dynamics of exciton separation and recombination [48–53]. As opposed to the case in M-MNP hybrids, the specific details of the chemical bond between the molecule and the NP are of paramount importance in controlling interfacial CT. This makes the study of these systems a complex problem in electronic structure because of the importance of many-body effects, the influence of electron correlation, and the difficulties involved in a full geometry optimization. In addition, any realistic treatment must include dynamical aspects in a time-dependent picture because of the different time scales involved in CT and exciton dynamics and an appropriate description of excited states that could be very relevant for all the processes mentioned above.

The inclusion of one or more electronic excited states, in addition to the ground state, has considerable methodological and practical implications. An accurate description of both fluorescence and Raman spectra depends on an appropriate characterization of the states involved in the process.

### *6.5.1 A Unified Description of Scattering and Fluorescence Processes in M-MNP Hybrids*

A model that has strongly influenced our thinking for this proposal was the one introduced in references [54, 55] for a system consisting of two metal NPs of radius R and a molecule, represented by a dipole at distances $d_1$and $d_2$ from the NPs, that is schematically presented in Fig. 1 (taken from Ref. [54]). The model considers two molecular electronic states which are approximated by two displaced harmonic oscillator potential energy curves corresponding to the ground (g) and excited (e) states. The inclusion of three vibrational states per harmonic oscillator for each harmonic potential completes the description of the quantum state manifold for this problem.

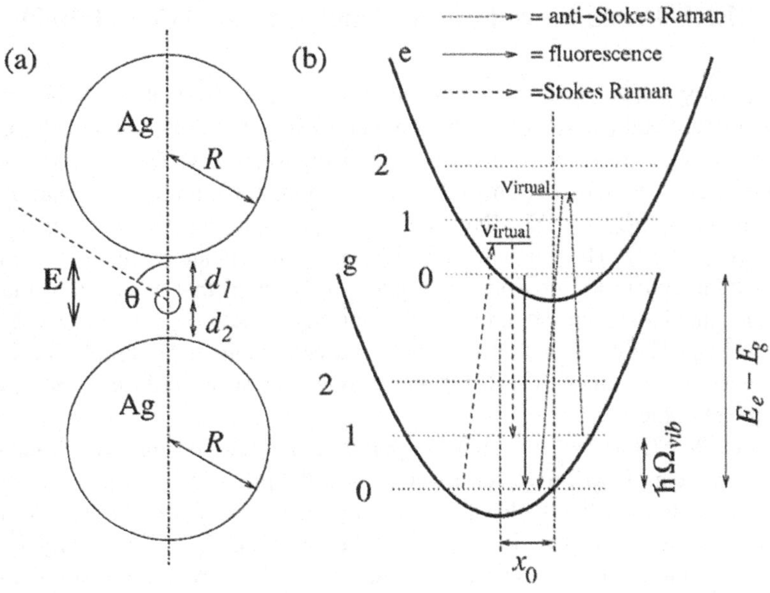


For the free molecule, the absorption and Raman cross sections can be calculated using first-order perturbation theory to be

$$\sigma_A = p_o^2 \frac{\Omega_L}{c\varepsilon_o} \sum_n \frac{\mathrm{h}\Gamma_{\mathrm{tot}} f(0,n) f(0,n)}{(\mathrm{h}\Omega_L - n\mathrm{h}\omega - \mathrm{h}\Omega_{ge})^2 + \mathrm{h}^2\Gamma_{tot}^2} \tag{6.25}$$

$$\sigma_R = p_o^4 \frac{\omega'^3 \Omega_L}{6\pi\varepsilon_o^2 c^4} \left| \sum_n \frac{\mathrm{h}\Gamma_{\mathrm{tot}} f(0,n) f(0,n)}{(\mathrm{h}\Omega_L - n\mathrm{h}\omega - \mathrm{h}\Omega_{ge})^2 + \mathrm{h}^2\Gamma_{tot}^2} \right|^2 \tag{6.26}$$

In the presence of the metal NPs, the two processes are interconnected via the time evolution of the effective dipole moment. The calculation of the total scattering cross section involving Raman and Rayleigh scattering and fluorescence can be computed in the following way. The emitted light intensity at position $\mathbf{r}_o$ is proportional to the field correlation function.

$$I_{em}(\mathbf{r}_o, t) = 2\varepsilon_o c \left\langle E_\theta^{(-)}(\mathbf{r}_o, t) E_\theta^{(+)}(\mathbf{r}_o, t) \right\rangle \tag{6.27}$$

Equation (6.27) can be transformed to the frequency domain using the Wiener-Khintchine to obtain the total power spectrum as

$$I_{em}(\mathbf{r}_o,\omega)=\frac{1}{\pi}\mathrm{Re}\int_0^{\infty}d\tau 2\varepsilon_o c\left\langle E_\theta^{(-)}(\mathbf{r}_o,0)E_\theta^{(+)}(\mathbf{r}_o,\tau)\right\rangle e^{i\omega t} \tag{6.28}$$

The model assumes that the electric fields are caused by the electric dipole moment $p(t)$ of the molecule which generates the radiated electric field $\vec{E}=\theta p e^{i(kr-\omega t)}\omega^2\sin\theta/\left(4\pi\varepsilon_o c^2 r\right)$. The double-differential scattering and fluorescence cross section is then obtained as

$$\frac{d^2\sigma}{d\Omega d(\hbar\omega)}=\frac{\omega^4\sin^2\theta}{I_{in}8\pi^3c^3\varepsilon_o\hbar}\mathrm{Re}\int_0^{\infty}d\tau e^{i\omega\tau}\left\langle p^{(-)}(0)p^{(+)}(\tau)\right\rangle \tag{6.29}$$

The next step in the calculation is to compute the dipole time-correlation function in Eq. (6.29), which in turns requires solving Liouville equation for the molecular density matrix. The total Hamiltonian for the system is written as

$$H=H_{\mathrm{mol}}+H'+H_{\mathrm{fluc}}+H_{\mathrm{env}}, \tag{6.30}$$

where $H_{\mathrm{mol}}$, $H'$, $H_{\mathrm{fluc}}$, and $H_{\mathrm{env}}$ are the molecular Hamiltonian, the molecule-field interaction, the molecule interaction with field vacuum fluctuations, and the interaction with any other material environment, respectively. Liouville equation can then be written as

$$i\frac{d\rho}{dt}=\frac{1}{\hbar}[H_{\mathrm{mol}}+H',\rho]+L_{tr}\rho+L_{ph}\rho \tag{6.31}$$

The first term in (6.31) describes the contribution from the molecular Hamiltonian and the interaction $H'$. The relaxation superoperators $L_{tr}$ describe the damping of the density matrix as a result of transitions caused by $H_{\mathrm{fluc}}$ that cause the spontaneous emission of photons and $L_{ph}$ describe relaxation and dephasing caused by the material environment, i.e., the coupling to the NPs. The explicit description of the relaxation superoperators in Reference [55] is achieved through the introduction of two sets of parameters $\Gamma_{kj}$ and $\gamma_{ph}$ representing the total decay rate from state k to state j and the dephasing rate, respectively. These parameters enter into Liouville equation in the following way:

$$L_{tr}\rho=-\sum_{kj}\frac{i\Gamma_{kj}}{2}\left[\sigma_{jk}\sigma_{kj}\rho+\rho\sigma_{jk}\sigma_{kj}-2\sigma_{kj}\rho\sigma_{jk}\right] \tag{6.32}$$

$$L_{ph}\rho_{jk}=-i\gamma_{ph}\rho_{kj} \tag{6.33}$$

with $\sigma_{kj}$ denoting a matrix with the only nonzero element "kj" equal to 1.

In general, a solution of the density matrix is not sufficient to calculate the dipole time-correlation function in (6.29); however, use of the Onsager-Lax quantum regression theorem, which hinges on the validity of the Markovian approximation, permits to calculate a two-time correlation function from a single-time correlation function, thereby providing all the necessary quantities to compute the absorption and scattering cross section (6.29).

### 6.5.2 *Density Matrix Treatment of Combined Instantaneous and Delayed Dissipation*

A second background ingredient that constitutes important guidance in our approach to this proposal is the work of Micha and coworkers [56–61] where a density matrix approach is used to simultaneously include relaxation, damping and dephasing processes that occur in different time scales, something that in their terminology is called instantaneous and delayed dissipation. This model assumes a separation of the entire system in a primary region that is described using a reduced density matrix (RDM) and a secondary region representing the environment. Fast dissipation is described by a Lindblad term associated with electronic transitions induced in the primary region by its interaction with the secondary one. The delayed dissipation is given by a time integral with a memory term derived from the time correlation of atomic displacements in the medium. The separation into instantaneous and delayed dissipation is based on the different time scales of electronic and vibrational transitions. The model has been successfully applied to a number of physically relevant situations involving the dynamics of electronically excited adsorbates on solid surfaces, a system where a realistic description of the chemical bond between the molecule and the surface is important, a situation reminiscent of the subject of our proposal.

The model is involved, but the basic equations can be understood using a simplified version of the quantum theory of relaxation. The total Hamiltonian for the system and the environment can be written in the usual way:

$$H = H_S + H_R + V \tag{6.34}$$

where (S) and (R) correspond to the primary and secondary region, respectively. Assuming that the interaction is switched at time $\underline{t} = 0$ and that prior to this S and R are uncorrelated, Liouville Eq. (6.31) can be written in the interaction picture as

$$\dot{\rho}_I(t) = -(i/\mathrm{h})[V_I(t), \rho_I(0)] - (1/\mathrm{h})^2 \int_0^t dt' \left[V_I(t), \left[V_I\left(t'\right), \rho_I\left(t'\right)\right]\right] \tag{6.35}$$

The RDM describing the system of interest S is the partial trace of the full density matrix with respect to the reservoir

$$\rho_{SI}(t) = tr_R \rho_I(t), \tag{6.36}$$

and its evolution equation can be found directly from ((6.35) as

$$\dot{\rho}_{SI}(t) = -(i/\mathrm{h}) tr_R [V_I(t), \rho_I(0)] - (1/\mathrm{h})^2 \int_0^t dt' \, tr_R \left[ V_I(t), \left[ V_I\left(t'\right), \rho_I\left(t'\right) \right] \right] \tag{6.37}$$

Two keys assumptions are commonly made to simplify Eq. (6.37):

(i) The reservoir is considered to be in thermal equilibrium at all times, that is,

$$\begin{aligned} \rho_I(t) &= \rho_{SI}(t) \rho_R(0) \\ \rho_R(0) &= \exp(-H_R/kT)/Z \end{aligned} \tag{6.38}$$

with Z the partition function.

(ii) The Markov approximation whereby memory effects are neglected. This amounts to making the replacement $\rho_{SI}(t') \rightarrow \rho_{SI}(t)$ in the integral in Eq. (6.35).

With these two approximations that introduce irreversibility into the dynamics of the system, the equation for the RDM is

$$\dot{\rho}_{SI}(t) = -(i/\mathrm{h}) tr_R [V_I(t), \rho_S(0) \rho_R(0)] - (1/\mathrm{h})^2 \int_0^t dt' \, tr_R \left[ V_I(t), \left[ V_I\left(t'\right), \rho_{SI}(t) \rho_R(0) \right] \right] \tag{6.39}$$

The last step is making the connection between the general equations of quantum relaxation and Micha et al.'s work which is to make an assumption about the specific form of the interaction V. This is assumed to be of the form

$$V_I(t) = \sum_i F_i(t) Q_i \tag{6.40}$$

where $F_i(t)$ and $Q_i(t)$ are reservoir- and system-only operators, respectively. Using the Markov approximation for the reservoir and including all the approximations mentioned so far, one can write Eq. (6.39) in such a way that all the information on the reservoir is contained in the time-correlation functions of the reservoirs:

$$i\mathrm{h} \frac{d\rho_{SI}}{dt} = -\frac{i}{\mathrm{h}} \sum_{ij} \int_0^\infty dt'' \left\{ \begin{array}{l} \left[ Q_i(t), Q_j\left(t - t''\right) \rho_{SI}(t) \right] \left\langle F_i\left(t''\right) F_j \right\rangle - \\ \left[ Q_i(t), \rho_{SI}(t) Q_j\left(t - t''\right) \right] \times \left\langle F_i F_j\left(t''\right) \right\rangle \end{array} \right\} \tag{6.41}$$

This can be written in a much more compact way by defining laxation superoperator $R$ such that

$$i\hbar \frac{d\rho_{SI}}{dt} = R\rho_{SI} \tag{6.42}$$

whose matrix components are directly related to the parameters $\Gamma_{lk}$ and $\gamma_{ph}$ and in Eqs. (6.32) and (6.33). Computing these parameters requires some explicit assumptions about the reservoir dynamics that must be tested for each specific system.

## 6.6 Conclusions and Final Remarks

We have described in this tutorial some of the basic physics underlying the design of molecular sensors using NPs with different electronic properties. In one case, plasmonic sensors are based on the optical response leading to changes on the plasmon resonance. In the other, the sensor is based on a molecular property, Raman spectrum, that is strongly enhanced due to interfacial charge transfer between the molecule and the NP. We presented very recent experimental results using plasmonic sensors that can be understood using a Mie theory-based model for the plasmonic response.

## References

1. Boisselier E, Astruc D (2009) Gold nanoparticles in nanomedicine: preparations, imaging, diagnostics, therapies and toxicity. Chem Soc Rev 38(6):1759–1782
2. Spackova B et al (2016) Optical biosensors based on Plasmonic nanostructures: a review. Proc IEEE 104(12):2380–2408
3. Lin HS, Carey JR (2014) The design and applications of nanoparticle coated microspheres in immunoassays. J Nanosci Nanotechnol 14(1):363–377
4. Odom TW, Schatz GC (2011) Introduction to Plasmonics. Chem Rev 111(6):3667–3668
5. Barnes WL, Dereux A, Ebbesen TW (2003) Surface plasmon subwavelength optics. Nature 424 (6950):824–830
6. Hao T, Riman RE (2006) Calculation of interparticle spacing in colloidal systems. J Colloid Interface Sci 297(1):374–377
7. Daniel MC, Astruc D (2004) Gold nanoparticles: assembly, supramolecular chemistry, quantum-size-related properties, and applications toward biology, catalysis, and nanotechnology. Chem Rev 104(1):293–346
8. Kuznetsov AI et al (2016) Optically resonant dielectric nanostructures. Science 354:6314
9. Nguyen VH, Nguyen BH (2015) Basics of quantum plasmonics. Adv Nat Sci Nanosci Nanotechnol 6(2):023001
10. Haiss W et al (2007) Determination of size and concentration of gold nanoparticles from UV-Vis spectra. Anal Chem 79(11):4215–4221
11. Khlebtsov NG et al (2003) A multilayer model for gold nanoparticle bioconjugates: application to study of gelatin and human IgG adsorption using extinction and light scattering spectra and the dynamic light scattering method. Colloid J 65(5):622–635
12. Dreaden EC et al (2012) The golden age: gold nanoparticles for biomedicine. Chem Soc Rev 41 (7):2740–2779

13. Brewer SH et al (2005) Probing BSA binding to citrate-coated gold nanoparticles and surfaces. Langmuir 21(20):9303–9307
14. Dominguez-Medina S et al (2012) In situ measurement of bovine serum albumin interaction with gold nanospheres. Langmuir 28(24):9131–9139
15. Dewi MR, Laufersky G, Nann T (2014) A highly efficient ligand exchange reaction on gold nanoparticles: preserving their size, shape and colloidal stability. RSC Adv 4(64):34217–34220
16. Ambrosi A et al (2007) Double-codified gold nanolabels for enhanced immunoanalysis. Anal Chem 79(14):5232–5240
17. Cliffel DE, Turner BN, Huffman BJ (2009) Nanoparticle-based biologic mimetics. Wiley Interdiscip Rev Nanomed Nanobiotechnol 1(1):47–59
18. Pamies R et al (2014) Aggregation behaviour of gold nanoparticles in saline aqueous media. J Nanopart Res 16(4):2376
19. Hohenester U, Trugler A (2012) MNPBEM: a Matlab toolbox for the simulation of plasmonic nanoparticles. Comput Phys Commun 183(2):370–381
20. de la Rica R, Stevens MM (2012) Plasmonic ELISA for the ultrasensitive detection of disease biomarkers with the naked eye. Nat Nanotechnol 7(12):821–824
21. Mani V et al (2009) Ultrasensitive immunosensor for cancer biomarker proteins using gold nanoparticle film electrodes and multienzyme-particle amplification. ACS Nano 3(3):585–594
22. Barnes WL, Dereux A, Ebbesen TW (2003) Surface plasmon subwavelength optics. Nature 424 (6950):824–830
23. Parveen S, Misra R, Sahoo SK (2012) Nanoparticles: a boon to drug delivery, therapeutics, diagnostics and imaging. Nanomed Nanotechnol Biol Med 8(2):147–166
24. Tarakeshwar P, Finkelstein-Shapiro D, Rajh T, Mujica V (2010) Quantum confinement effects on the surface enhanced Raman spectra of. Int J Quantum Chem, to be published.
25. He L, Musick MD, Nicewarner SR, Salinas FG, Benkovic SJ, Natan MJ, Keating CD (September 2000) Colloidal Au-enhanced surface Plasmon resonance for ultrasensitive detection of DNA hybridization. J Am Chem Soc 122(38):9071–9077
26. Katherine A Willets and Richard P Van Duyne (January 2007) Localized surface plasmon resonance spectroscopy and sensing. Annu Rev Phys Chem 58 267–297
27. Anker JN, Hall WP, Lyandres O, Shah NC, Zhao J, Van Duyne RP (June 2008) Biosensing withplasmonicnanosensors. Nat Mater 7(6):442–453
28. Stewart ME, Anderton CR, Thompson LB, Maria J, Gray SK, Rogers JA, Nuzzo RG (February 2008) Nanostructured plasmonic sensors. Chem Rev 108(2):494–521
29. Willner I, Willner B, Katz E (January 2007) Biomolecule-nanoparticle hybrid systems for bioelectronic applications. Bioelectrochem (Amsterdam, Netherlands) 70(1):2–11
30. O'Regan B, Grätzel M (October 1991) A low-cost, high-efficiency solar cell based on dyesensitized colloidal $TiO_2$ films. Nature 353(6346):737–740
31. Nazeeruddin MK, Kay A, Rodicio I, Humphry-Baker R, Mueller E, Liska P, Vlachopoulos N, Grätzel M (July 1993) Conversion of light to electricity by cis-X2bis (2,2′-bipyridyl4,4-′-dicarboxylate) ruthenium (II) charge-transfer sensitizers (X = Cl-, Br-, I-, CN-, and SCN) on nanocrystalline titanium dioxide electrodes. J Am Chem Soc 115(14):6382–6390
32. Hagfeldt A, Graetzel M (June 1995) Light-induced redox reactions in nanocrystalline systems. Chem Rev 95(1):49
33. Hagfeldt A, Grätzel M (May 2000) Molecular photovoltaics. Acc Chem Res 33(5):269–277
34. Schnadt J, Bruhwiler PA, Patthey L, O'Shea JN, Södergren, S, Odelius M, Ahuja R, Karis O, Bässler M, Persson P, Others (2002) Experimental evidence for sub-3-fs charge transfer from an aromatic adsorbate to a semiconductor. Nature 418(6898):620–623
35. Grätzel M (February 2003) Solar cells to dye for. Nature 421:586–587
36. Nie S (February 1997) Probing single molecules and single nanoparticles by surface-enhanced Raman scattering. Science 275(5303):1102–1106
37. Kneipp K, Wang Y, Kneipp H, Perelman L, Itzkan I, Dasari R, Feld M (March 1997) Single molecule detection using surface-enhanced Raman scattering (SERS). Phys Rev Lett 78 (9):1667–1670

38. E C Le R, Meyer M, Etchegoin PG (February 2006) Proof of single-molecule sensitivity in surface enhanced Raman scattering (SERS) by means of a two-analyte technique. J Phys Chem B 110(4):1944–1948
39. Hongxing X, Bjerneld E, Käll M, Börjesson L (November 1999) Spectroscopy of single hemoglobin molecules by surface enhanced Raman scattering. Phys Rev Lett 83 (21):4357–4360
40. Hutter E, Fendler JH (October 2004) Exploitation of localized surface plasmon resonance. Adv Mater 16(19):1685–1706
41. Gratzel M (June 2004) Conversion of sunlight to electric power by nanocrystalline dye-sensitized solar cells*1. J Photochem Photobiol A Chem 164(1–3):3–14
42. Law M, Greene LE, Johnson JC, Saykally R, Yang P (June 2005) Nanowire dye-sensitizedsolarcells. Nat Mater 4(6):455–459
43. Wasielewski MR (May 1992) Photo induced electron transfer in supramolecular systems for artificial photosynthesis. Chem Rev 92(3):435–461
44. Bard AJ, Fox MA (March 1995) Artificial photosynthesis: solar splitting of water to hydrogen and oxygen. Acc Chem Res 28(3):141–145
45. Kay A, Graetzel M (June 1993) Artificial photosynthesis. 1. Photosensitization of titania solar cells with chlorophyll derivatives and related natural porphyrins. J Phys Chem 97 (23):6272–6277
46. Kay A, Humphry-Baker R, Graetzel M (January 1994) Artificial photosynthesis. 2. Investigations on the mechanism of photosensitization of Nanocrystalline $TiO_2$ solar cells by chlorophyll derivatives. J Phys Chem 98(3):952–959
47. Asahi R, Morikawa T, Ohwaki T, Aoki K, Taga Y (July 2001). Visible-light photocatalysis in nitrogen doped titanium oxides. Science (NewYork, NY) 293(5528):269–271.
48. Zhang Z, Wang C-C, Zakaria R, Ying JY (December 1998) Role of particle size in nanocrystalline $TiO_2$-based photocatalysts. J Phys Chem B 102(52):10871–10878
49. Ramakrishna S, Willig F, May V (2001) Theory of ultrafast photo induced heterogeneous electron transfer: decay of vibrational coherence into a finite electronic vibrational quasicontinuum. J Chem Phys 115(6):2743
50. Benkö G, Kallioinen J, Korppi-Tommola JEI, Yartsev AP, Sundström V (2002) Photo induced ultrafast dye-to-semiconductor electron injection from nonthermalized and thermalized donor states. J Am Chem Soc 124(3):489–493
51. Stier W, Prezhdo OV (August 2002) Nonadiabatic molecular dynamics simulation of light induced electron transfer from an anchored molecular electron donor to a semiconductor acceptor. J Phys Chem B 106(33):8047–8054
52. Duncan WR, Stier WM, Prezhdo OV (June 2005) Ab initio nonadiabatic molecular dynamics of the ultrafast electron injection across the alizarin-$TiO_2$ interface. J Am Chem Soc 127 (21):7941–7951
53. Guo Z, Liang WZ, Yi Z, Chen GH (October 2008) Real-time propagation of the reduced one-electron density matrix in atom-centered orbitals: application to Electron injection dynamics in dye-sensitized $TiO_2$ clusters. J Phys Chem C 112(42):16655–16662
54. Le Ru EC, Etchegoin PG (2008) Principles of surface-enhanced Raman spectroscopy: and related Plasmonic effects. ElsevierScience Ltd, Amsterdam
55. Hongxing X, Wang X-H, Martin Persson HX, Käll M, Johansson P (December 2004) Unified treatment of fluorescence and Raman scattering processes near metal surfaces. Phys Rev Lett 93 (24):1–4
56. Johansson P, Hongxing X, Käll M (July 2005) Surface-enhanced Raman scattering and fluorescence near metal nanoparticles. Phys Rev B 72(3):1–17
57. Salam A, Micha DA (1999) Nonlinear optical response of metal surfaces with adsorbed molecules. Int J Quantum Chem 75(4–5):429–439
58. Zhigang Yi DA (1999) Micha, and James Sund. Density matrix theory and calculations of nonlinear yields of CO photodesorbed from Cu (001) by light pulses. J Chem Phys 110 (21):10562

59. Micha DA, Santana A, Salam A (2002) Nonlinear optical response and yield in the femtosecond photodesorption of CO from the Cu (001) surface: a density matrix treatment. J Chem Phys 116 (12):5173
60. Leathers AS, Micha DA, Kilin DS (March 2010) Direct and indirect electron transfer at a semiconductor surface with an adsorbate: theory and application to Ag3Si (111):H. J Chem Phys 132(11):114–702
61. Kilin DS, Micha DA (2010) Modeling the photovoltage of doped Si surfaces. J Phys Chem C, to appear, 115(3):797–858,

# Chapter 7
# From Micro- to Nano-objects and from Giga- to Terahertz Frequency Range: Quasi-optical Resonant Diagnostics of Objects

**Grigory Dunaevsky and Igor Dorofeev**

**Abstract** A highly sensitive method of contactless microwave diagnostics is considered that allows to examine the properties of small objects of different shapes and permittivities and small local inhomogeneities of artificial (composite) and natural materials. The method is based on multiple interactions of radio waves with an examined object in an open quasi-optical resonator. It can be used in a wide frequency range, from units of gigahertz to several tens of terahertz, unachievable by other methods.

The capabilities of the method are illustrated by detection of local nanostructured inhomogeneities of super thin threadlike wires under a glass insulation layer, measurement of polarizabilities of small spherical aerogels from carbon nanotubes, technological diagnostics of thin film thickness, and measurement of inhomogeneities of the reflection coefficient of sheet composites.

## 7.1 Introduction

In all stages of the development of radio electronics, simultaneously with mastering of frequencies in each range of the radio wave spectrum, materials possessing desired electromagnetic characteristic in the given frequency range have been synthesized. Accordingly, in each stage of the development, methods for measuring these characteristics at these frequencies have also been developed. In this work, we consider the method of measurements which in our opinion is urgent nowadays. This method is aimed at radio wave measurements and investigation of small-sized objects actively used as components of new composites and nanomaterials. It is important that this method is applicable in a very wide frequency range of the electromagnetic spectrum, from several gigahertz to several tens of terahertz.

G. Dunaevsky (✉) · I. Dorofeev
Department of Radioelectronics, Tomsk State University, Tomsk, Russia
e-mail: proecs@mail.tsu.ru



S. M. Goodnick et al. (eds.), *Semiconductor Nanotechnology*, Nanostructure Science and Technology, https://doi.org/10.1007/978-3-319-91896-9_7

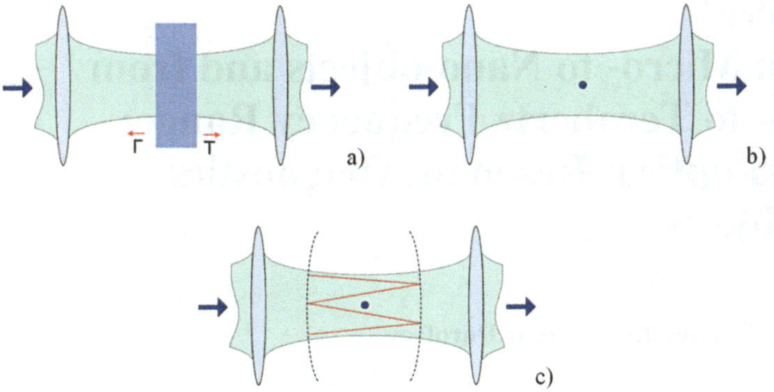


**Fig. 7.1** Examples of single (**a**) and (**b**) and multiple (**c**) interaction of electromagnetic radiation with the examined object

Traditional scheme of measuring radio wave parameters of the examined material sample involves its arrangement in an electromagnetic wave field, for example, as shown in Fig. 7.1a, between the lens that sends radiation to the object and the lens that collects the wave transmitted through it; in this case, changes of the amplitude and phase of the wave transmitted through the object and (or) the wave reflected from it are measured. In case the object is small, as shown in Fig. 7.1b, changes it creates in the beam parameters become weakly noticeable for detection of both reflected and transmitted waves against the background noise. The efficiency of wave interaction with the object can be increased in all such cases, if the wave passes through the object more than once, for example, as shown in Fig. 7.1c, where semitransparent reflectors are arranged to the right and to the left of the examined object.

The method we suggest to consider in details is based on multiple interaction of the probing wave with the examined object in the so-called open resonator shown in Fig. 7.1c and formed by plane or more often concave mirrors [1]. In optics this design is known as the Fabry–Perot interferometer, and the resonant phenomena in these structures formed by two reflecting elements are called the Fabry–Perot resonances. Nowadays fundamental investigation of these phenomena is still continued [2] since the Fabry–Perot resonances provide the basis for continuously developing laser technology. Results of investigation of the Fabry–Perot resonances are important for a description of both electromagnetic [3] and acoustic [4] waves and are also used for investigation and design of novel metamaterials [5], nano-objects [6], and graphene layers [7].

The open radio wave resonator of the Fabry–Perot type has unique properties. The simplest design consisting of two coaxial concave metal reflectors whose apertures significantly exceed the wavelength has equidistant spectrum of resonant frequencies with stable Q-factor that can reach several tens and even hundreds of thousands. Technically it is possible to separate natural oscillations with high Q-factor in the same resonator in the continuous wavelength range: from centimeter to optical one.

Such resonator has clear advantages for diagnostics because it provides free and arbitrary arrangement of the examined object between the reflectors. In addition to convenience, this can be of crucial importance when the open resonator is used as a sensor in systems of remote control of moving objects. These advantages result in the application of the quasi-optical resonator for measurements of the material parameters practically since the first years of its development [8] especially at gigahertz frequencies where measurements in volume resonators have shown to be highly effective.

However, despite several advantages in many aspects, the application of open resonators, including the terahertz frequency range, is still limited. Thus in one of the recent sufficiently detailed reviews of the current state of research in the terahertz frequency range [9], only free space and waveguide methods have been considered as the main methods of terahertz metrology for two essential reasons. The first is technical: because of high quality of open resonators, high requirements are imposed on measurement technique used to fix changes in the Q-factor and resonant frequency shift. However, these problems are being gradually solved; see [10–13].

The second more essential reason of their limited application is that diffraction phenomena are inevitable in the open resonator with an object; their consideration creates additional difficulties in the interpretation of measurements. Therefore, from the first works devoted to the application of the open resonator in metrology of dielectrics [8] to the recent ones [14], large-area objects with plane parallel geometry, for which diffraction effects can be neglected, are mainly investigated in open resonators. However, as already indicated above, the shapes of examined objects are actually much more diverse.

Therefore, below we focus on the possibility of investigating small-sized objects in the open resonator considering diffraction effects in the corresponding mathematical models and demonstrate the feasibility of diagnostics of small-sized threadlike and spherical samples. We will see how the allowance for diffraction makes it possible to solve the problem of diagnostics of moving objects and reflections from small-sized fragments of large composite samples with losses.

The fact that the greater part of experimental confirmation of the suggested techniques was obtained in giga- rather than terahertz frequency range should not embarrass; on the contrary, longer wavelengths allowed us to compare more carefully the results of measurements and calculations. Since the elemental base of the terahertz frequency range is being rapidly developed, the above-indicated technical problems in this frequency range will be overcome; the suggested techniques can be employed for diagnostics of nano-objects, nanostructural macroobjects, and nanocomposites at all frequencies from the terahertz spectrum.

## 7.2 Quasi-optical Open Resonator and Its Special Features

As already indicated above, the simplest open radio wave resonator can be formed by plane parallel mirrors, but the open resonator formed by concave spherical reflectors is much more convenient for measurements. Such resonator with correctly chosen geometry is more stable to imbalance between mirrors and possesses much higher quality.

For the resonator formed by concave mirrors with radius $R_m$ spaced at $L_{res}$, the resonant frequencies are given by the formula [15]:

$$f = \frac{c}{2L_{res}}\left[q + 1 + \frac{1}{\pi}(2m + n + 1)\arccos\left(1 - \frac{L_{res}}{R_m}\right)\right], \tag{7.1}$$

where $c$ is the velocity of light and $m$, $n$, and $q$ are indices for the $TEM_{mnq}$ mode. From this formula, it can be seen that at $m = 0$ and $n = 0$, that is, without higher harmonic oscillations, the resonant frequency spectrum represents equidistant sequence with resonant frequencies spaced at $c/2\ L$. For mirrors spaced at about 10 cm, the frequency interval between the neighboring resonances will be 1.5 GHz. With such *step* size, the spectral characteristics of objects placed in the resonator can be investigated without resonator retuning.

Placing of the examined object in the resonator leads to shift of resonant curves, and as demonstrated below, the parameters of the examined sample can be estimated from the observed shift of the resonant frequency from that given by Eq. (7.1).

The loaded quality factor of the OR depends on its length $L_{res}$ and power losses for one circular beam pass [15]:

$$Q = \frac{f}{2\Delta f} = \frac{2k_1 L_{res}}{\eta_\Sigma}, \tag{7.2}$$

$$\eta_\Sigma = \eta_1^R + \eta_2^R + \eta_1^D + \eta_2^D + \eta_1^C + \eta_2^C + \eta_1^{DC} + \eta_2^{DC} + \eta_0. \tag{7.3}$$

Here $k_1 = \omega/c, \omega = 2\pi f$ is the circular frequency, and $2\Delta f$ is the width of the observed resonant curve. Here $\eta_1^R$ and $\eta_2^R$ are losses due to the surface impedance of the mirrors, $\eta_1^D$ and $\eta_2^D$ are losses due to diffraction on the mirrors, $\eta_1^C$ and $\eta_2^C$ are losses due to radiation in external circuits, $\eta_1^{DC}$ and $\eta_2^{DC}$ are losses due to diffraction on the coupling elements, and $\eta_x$ are losses on the measurable object.

During measurements, all types of losses in the resonator, except losses on the examined sample, are assumed to be constant. If the Q-factor decreases when the sample is placed in the resonator that is manifested through the increase by $2\Delta f$ of the width of the resonant curve, the losses $\eta_x$ introduced by the object can easily be determined from line broadening (Eqs. (7.2) and (7.3)). It should be noted that, despite of a seemingly large number of intrinsic resonator losses against the background of which a small object must be detected, these losses are minimized by careful manufacture of mirrors, correct choice of their apertures, and optimal resonator coupling with the external path (considered separately below).

The examined object is most often placed in the beam cross section where the wave is closest to the plane wave. For the above described resonator formed by identical concave mirrors, the plane wave front is formed in the center where the beam radius is [15]

$$w_0 = \sqrt{\frac{cL}{\omega}}\sqrt[4]{\frac{2R_m}{L} - 1}. \tag{7.4}$$

In this case, the radius of the field spot on the mirrors is described by the formula:

$$w_m = \sqrt{\frac{2cL}{\omega}}\frac{1}{\sqrt[4]{\frac{2R_m}{L} - 1}}. \tag{7.5}$$

Comparing Eqs. (7.4) and (7.5), we see that the beam radius $w_0$ in the resonator center can be minimized by approaching the distance $L$ between the mirrors to the doubled mirror radius $2R_m$; however, in this case the radius $w_m$ of the field spot on the mirrors will also increase, thereby increasing the diffraction losses $\eta_1^D$ and $\eta_2^D$ on the mirrors, and the resonator Q-factor will decrease.

Thus, placing the examined object between the resonator mirrors, we change the resonance conditions, that is, we change the resonant frequency and also introduce additional energy losses, that is, reduce the resonator Q-factor. Moreover, the objects that during single interaction introduced insignificant amplitude and (or) phase perturbations in the quasi-optical beam, being placed in the OR field, multiply and interact with the beam and produce, as demonstrated below, noticeable and well-measurable resonant frequency shifts and change the Q-factor of the resonator.

Below we consider the special features of investigation of objects with different sizes and shapes made of materials with various permittivity and conductivity.

## 7.3 Superthin Wire in an Open Resonator

It is expedient to investigate and to control nondestructively by the OR method one of the extraordinary objects – superthin extended micro- or nanowire. An example of such superthin wire is a glass-coated cast microwire produced by the Ulitovsky–Tailor method [16]; its diameter can be of the order of 100 nm [17]. Unlike nanowires used in microelectronics and deposited on a substrate [18], the microwire can exist without substrate, since it is covered by glass insulation that ensures sufficient mechanical strength. Between the glass insulation and the conductive core, there is the nanostructured transitive layer formed due to glass diffusion into metal and due to oxides [19]. Moreover, the microwire is often nonuniform because of complexity of the production process. Its characteristic nonuniformities have, as a rule, nanometer sizes.

An interesting feature of the microwire is the conductivity size effect according to which the conductivity of the microwire material differs considerably from the conductivity of the bulk material and depends on the wire radius. Some attempts were undertaken to construct a model of microwire conductivity [20–23]. Thus, for all these models, the empirical data are required which can be obtained only experimentally.

At high frequencies, the microwire possesses a number of unique properties from the viewpoint of its interaction with electromagnetic field. Therefore, nowadays its main application field is connected with its use as one of the components of composite materials rather than with signal transfer or processing. For example, microwire application is very efficient in the materials used to protect against radio wave radiation, since one of its special features is high dissipation ability [24]. This is caused by the fact that sufficiently weak skin effect can be observed in the microwire with high conductivity of the material of this wire. When the skin layer is comparable to the wire radius, the electromagnetic field penetrates completely into the wire without reflection and is dissipated as heat. The nanostructured transitive layer is in this case additional matching element providing more efficient energy transfer. It is clear that the microwire diagnostics by the radio wave methods should be performed at frequencies at which such composite material will subsequently be used.

As the volume of the superthin wire is very small, it is almost impossible to fix the result of its single interaction with a quasi-optical electromagnetic beam. When placed in the resonator field, the microwire influences noticeably both the Q-factor and resonant frequencies, and these changes can be measured. However, the mechanism of interaction of the OR field with the wire is not limited by internal dissipation and corresponding thermal losses in the wire volume; diffraction is also important here. Thus, the observed losses of wire in the OR are the sum of losses of these two types – thermal and diffraction ones.

Mathematical simulation gives a more detailed pattern of interaction of the electromagnetic beam of the OR field with an extended wire.

Let us consider a superthin wire placed into the OR perpendicular to the longitudinal resonator axis (Fig. 7.2) and parallel to the electric field vector $E$ of the resonator. We assume that the fundamental oscillation mode is excited in the OR with an odd longitudinal index, plane phase front, and field intensity distribution over the beam cross section close to Gaussian distribution.

According to [25], the OR resonant frequency shift is given by the following expression:

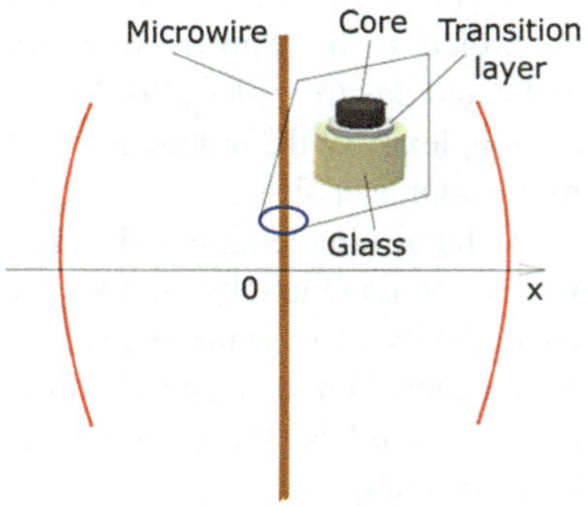


**Fig. 7.2** Glass-coated microwire arranged in the open resonator

$$\Delta f = \frac{i \int\limits_{v} jEdv}{4\pi N}, \tag{7.6}$$

where the function $j$ describes the current induced in the object and the function $E$ describes the field of OR natural oscillations. Integration formally should be carried out over the entire resonator volume and even over the surrounding space (the resonator is open). However, the product of functions being integrated is nonzero only in the microwire volume.

The quantity $N$ in the denominator is the so-called norm of the corresponding oscillation mode. For a fundamental mode, its value can be calculated from the formula [25]:

$$N = \frac{\pi}{4} w_0^2 \varepsilon_0 L_{\text{res}} E_0^2. \tag{7.7}$$

Here $\varepsilon_0$ is the permittivity of free space, and $E_0$ is the electric field in an antinode. This value is comparable with the energy stored in the resonator, and for further reasoning, we assume that controllable objects placed into the resonator are small in comparison with the resonator volume, so that the $N$ value changes negligibly when objects are placed into the resonator.

Generally, $\Delta f$ is a complex value; its real part $\delta f'$ describes the resonant frequency shift, and the doubled imaginary part $2\Delta f''$ describes resonant curve broadening caused by losses introduced by the object placed into the open resonator. Taking into account that the microwire introduces losses of two types – thermal and diffraction ones – we can write

$$2\Delta f''_W = 2\Delta f''_T + 2\Delta f''_D. \tag{7.8}$$

The current $j$ in the wire induced by the resonant oscillation field can be found by solving the corresponding boundary problem on the diffraction of the Gaussian OR beam on a thin cylinder. As a result of solution, the following expression for the resonant frequency shift was obtained from Eq. (7.6):

$$\delta f' = \text{Re}\left[\frac{2r_0}{\sqrt{2\pi} w_1 w_0 \varepsilon_0 L}\left(\sum_{n=-\infty}^{\infty} \left(J'_n(k_1 r_0) + A_n^{(1)} H_n^{(2)'}(k_1 r_0)\right)\right)\right], \tag{7.9}$$

where $w_1 = \sqrt{\mu_0/\varepsilon_0}$, $J_n$ is the Bessel function, and $H_n^{(2)}$ is the Hankel function. The coefficient

$$A_n = \frac{w_1 J'_n(k_2 r_0) J_n(k_1 r_0) - w_2 J_n(k_2 r_0) J'_n(k_1 r_0)}{w_2 J_n(k_2 r_0) H_n^{(2)'}(k_1 r_0) - w_1 J'_n(k_2 r_0) H_n^{(2)}(k_1 r_0)}, \tag{7.10}$$

$k_2 = k_1\sqrt{\varepsilon_2 \mu_2}$, $w_2 = w_1\sqrt{\mu_2/\varepsilon_2}$, $\varepsilon_2$ is the dielectric permittivity, and $\mu_2$ is the magnetic permeability of the wire.

To determine a relationship between the losses in the wire volume (thermal) and the diffraction losses, their separate consideration is required [26]. The power scattered in the wire volume on its conductivity $\sigma$ is determined by integration of the internal electric field $E^{\mathrm{int}}$ in the wire over its volume:

$$P_T = \frac{\sigma}{2} \int_v E^{\mathrm{int}} E^{\mathrm{int}*} \mathrm{d}v. \tag{7.11}$$

The power of the diffraction losses is calculated by integration of the diffracted fields $E^D$ and $H^D$ over the far field zone:

$$P_D = -\frac{1}{2} \int_S E^D H^{D*} \mathrm{d}S. \tag{7.12}$$

Total broadening of resonant curve (7.8) assumes the form:

$$2\Delta f''_W = 2\Delta f''_T + 2\Delta f''_D = \frac{2f}{k_1 L_{\mathrm{res}}} \frac{(P_T + P_D)}{P_{\mathrm{inc}}}, \tag{7.13}$$

where $P_{\mathrm{inc}} = \dfrac{\pi w_0^2 E_0^2}{4 w_1}$ is the incident beam power.

We now illustrate the behavior of the resonant frequency of the resonator and losses of various types attendant to changes in sizes and conductivity of the superthin cylindrical object by a numerical model. Let us introduce normalized quantities $Df_T = 2\Delta f''_T w_0 L$ and $Df_D = 2\Delta f''_D w_0 L$. We take the frequency equal to 10 GHz and the conductivity of the cylinder material equal to $10^7$ Sm/m. Figure 7.3 shows the plot of the dependence of $Df_T$ on the wire radius normalized by the skin-layer depth $\delta$.

As can be seen from this figure, a sharp increase in thermal losses is first observed. This is due to the fact that with increase in the wire diameter, the volume of wire interaction with the field increases, whereas the high electric field strength on the wire surface remains unchanged. This is the zone of the weak skin effect facilitating field penetration into the wire volume and providing efficient energy dissipation in the volume.

Then the phenomenon is observed, at first sight anomalous, namely, with increasing wire diameter, the losses in it do not increase, but, on the contrary, decrease! This effect is explained by the fact that first, with further increase in the wire diameter, the skin layer is formed that prevents the field penetration into the wire volume. Second, with increasing wire diameter, its resistivity decreases, and the electric field strength in its vicinity in the resonator decreases. As a result, the energy flux into the wire decreases, and the thermal losses also decrease. The higher the conductivity of the wire material, the faster this effect is realized.

The plot of $Df_D$ that describes the dependence of the diffraction losses on the normalized wire radius is shown in Fig. 7.4.

The diffraction losses in the initial segment of this dependence are small and further increase with the microwire diameter.

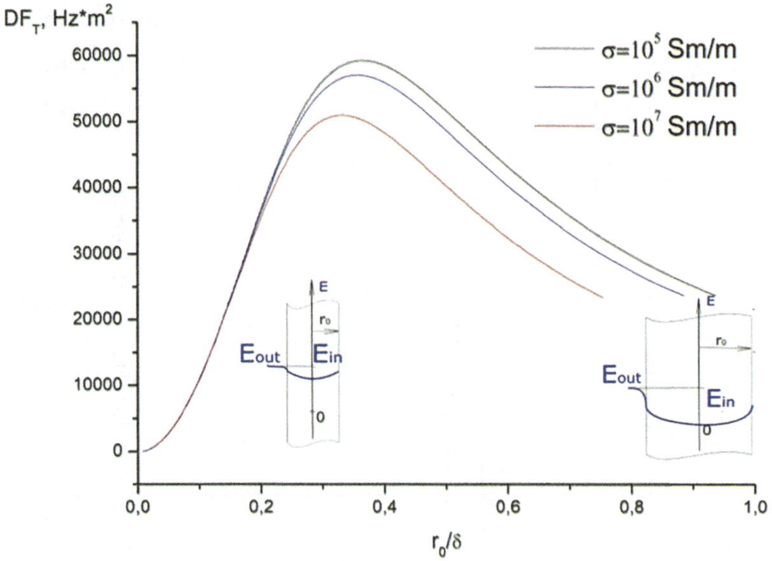


**Fig. 7.3** Thermal losses in the volume of a superthin wire depending on the wire radius normalized by the skin-layer depth for the indicated values of the bulk conductivity of metal

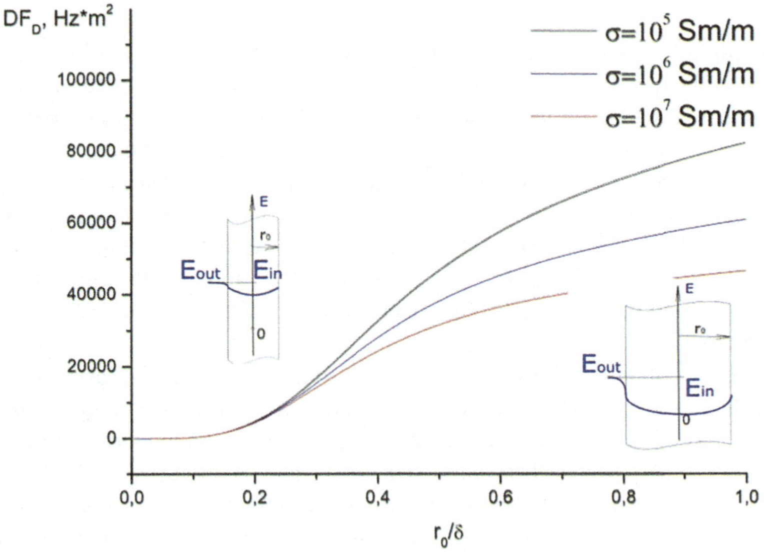


**Fig. 7.4** Diffraction losses introduced in the resonator by a superthin wire depending on its radius reduced to the skin-layer depth

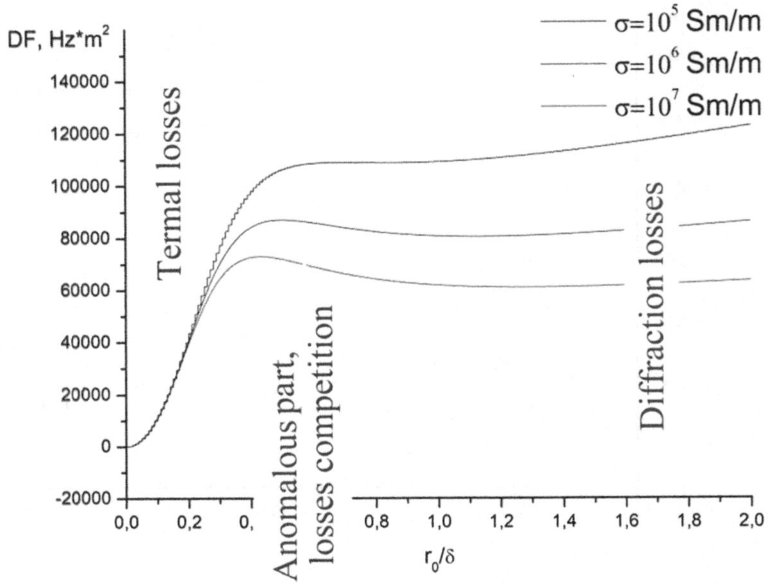


**Fig. 7.5** Total losses introduced by a superthin wire placed into the open resonator depending on the wire radius reduced to the skin-layer depth for the indicated values of the bulk conductivity

The plot of the total losses described by the *Df* value is shown in Fig. 7.5.

As can be seen from Fig. 7.5, the dependence of losses introduced in the open resonator by the wire on its diameter has some characteristic sections. In the initial section of the resonant curve for the resonator, the broadening is mainly determined by losses in the wire volume, whereas the diffraction losses here are small. The second section is anomalous in character; the Q-factor of the resonator in this section increases with the wire diameter. Here the two types of losses compete, the thermal losses decrease, and the diffraction losses increase. Finally, the third section is characterized by the slow logarithmic increase of the diffraction losses and insignificant thermal losses.

The results obtained allow us to interpret correctly the data of measuring these or other microwire parameters based on registration of changes in the Q-factor of the OR. Two sections are seen that allow contactless measurements of the microwire parameters to be performed. In the first section, the skin effect is weak, $2r_0 < 0.8\delta$, and the microwire diameter can be measured contactless here based on the increase of the losses. In the second section, the strong skin effect is observed, where the losses in the volume cease to play a significant role, and the diffraction losses prevail that change much more slowly with the wire radius.

As follows from the results of calculations performed, the efficient thermal dissipation and insignificant reradiation provide wires for which the condition $2r_0 < 0.8\delta$ is satisfied.

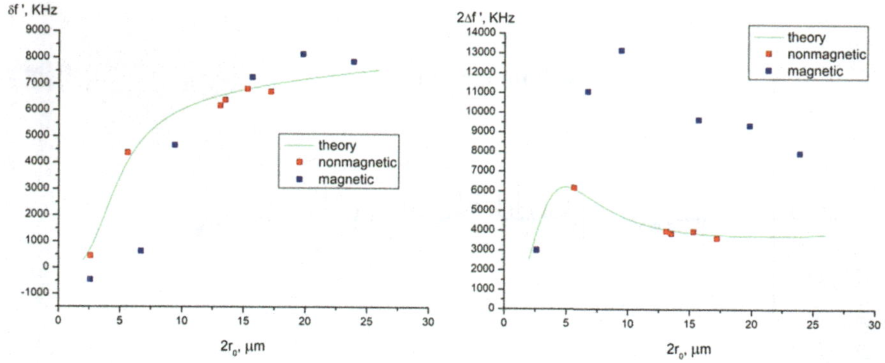


**Fig. 7.6** Open resonator with microwires at 9.38 GHz. Solid curves show results of calculations, blue symbols are for magnetic wires, and red curves are for nonmagnetic wires

The technologies of production of microwires possessing magnetic properties have been developed recently [16]. Investigations have shown that magnetization of such microwire is complex in character, and in this process, the anisotropy of the magnetic properties is manifested [27]. Nevertheless, the dissipative properties of a microwire can be improved at the expense of magnetic losses.

By way of illustration of operation of the model for a nonmagnetic microwire and influence of magnetic losses, Fig. 7.6 shows the calculated and experimental curves of the resonant frequency shift and broadening of the resonant curve for the open resonator at a frequency of 9.38 GHz [28].

As can be seen from Fig. 7.6, the magnetic microwire introduces losses in the open resonator that more than twice exceed those for the nonmagnetic microwire, except the case of the thinnest samples (with core 2.6 μm in diameter).

As a whole, this model adequately describes the physical pattern of electromagnetic radiation interaction with superthin wires.

The microwire production technology used nowadays does not allow stable parameters to be obtained even within one extended sample (one coil); therefore, the possibility of continuous contactless diagnostics of a moving wire becomes important. For this purpose, the device has been developed in which the open resonator acted as a detector [29]. The output signal of this detector represents the signal from the generator that has passed through the OR with a moving wire. To reduce the effect of wire vibration during its passage through the resonator, the wire was tilted to the plane parallel to the mirror apertures rather than oriented exactly in this plane. With the help of this device, very small (comparable to the wire diameter) local inhomogeneities of glass-coated microwires were detected. An example of the recorded detector signal is shown in Fig. 7.7. Three inhomogeneities were detected for the 1-m wire segment during pulling of the wire 5.6 μm in diameter; an analysis of these microwire segments using a microscope confirmed the presence of local extraneous inclusions under the glass insulation layer [30].

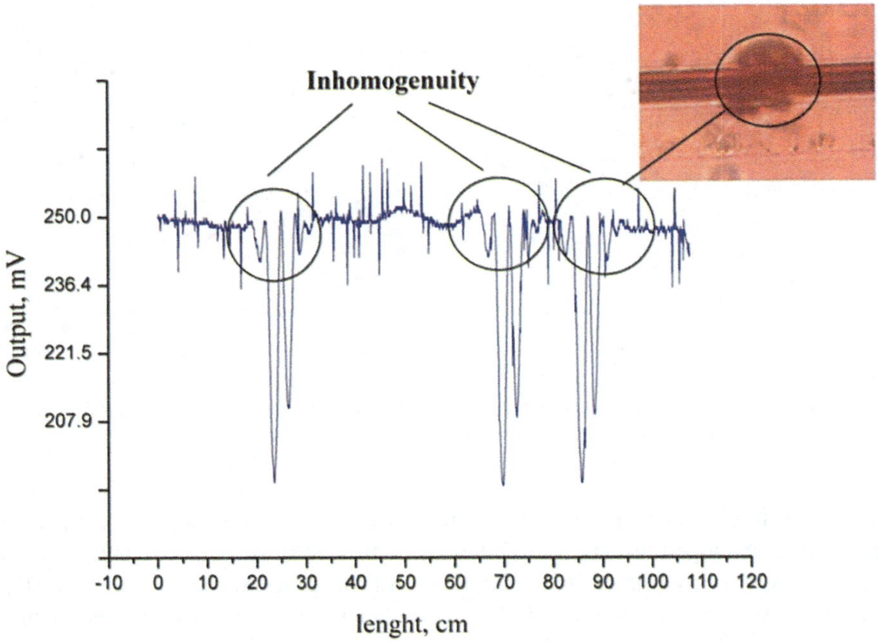


**Fig. 7.7** Inhomogeneities of a microwire

Thus, the OR can be used for diagnostics of superthin wires to perform contactless measurements of the radius and conductivity of nanowires and to reveal their local inhomogeneities including those hidden under the insulating dielectric layer.

## 7.4 Segment of a Superthin Wire

In composites, segments of microwires of small lengths are often used as components [31]. Of interest is a study of the properties of such elements using the OR. We consider that the length of a wire segment does not exceed the beam diameter in the OR. To use the same approach as for an extended wire, solutions of the problem of diffraction on a wire of finite length must be found. There are some methods of solving this problem. With the help of the approximate analytical methods described in [32], the current induced by the OR field in the microwire segment of length $2\,l$ was obtained based on which, with allowance for Eq. (7.6), we derive for the resonant frequency shift [33]:

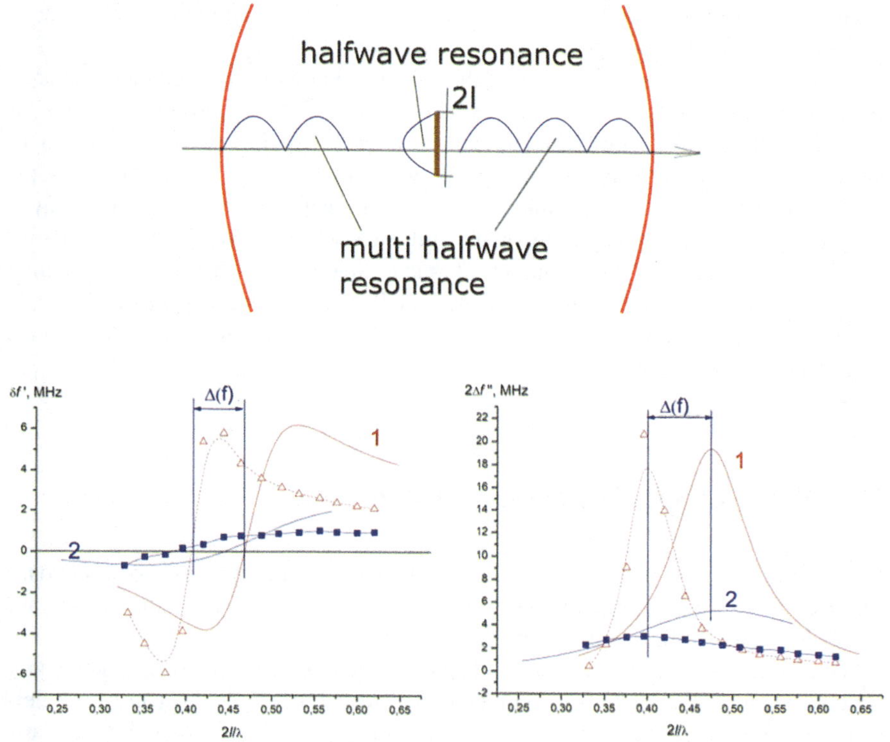


**Fig. 7.8** Open resonator with short microwires. Solid curves show the results of calculations by Eq. (7.14): *1* is for copper microwire with 5.6 μm core, and *2* is for resistive alloy with 5.6 μm core. Symbols are for experimental data. Open triangles are for copper, and closed squares are for a resistive alloy

$$\Delta f = \frac{2\alpha f \int_{-l}^{l} \left[\frac{\cos(k_2 s) - \cos(k_2 l)}{\cos(k_2 l) + \alpha P_L^s}\right] \mathrm{d}s}{\pi k_1 k_2 \varepsilon_0 w_0^2 L_{\mathrm{res}}}, \tag{7.14}$$

where

$$P_L^s = \int_{-l}^{l} \frac{\exp\left(-ik_1\sqrt{(s-l)^2 + r_0^2}\right)}{\sqrt{(s-l)^2 + r_0^2}} \cos(k_1 s)\mathrm{d}s. \tag{7.15}$$

Here $\alpha = \frac{1}{2\ln\left(\frac{r}{2l}\right)}$ is a small parameter. In [33], this model was tested for non-insulated wires (produced by the spinneret technology without insulating glass layer). The plots in Fig. 7.8 show the measured characteristics of the open resonator with samples 5.6 μm in diameter made of copper and alloy with material conductivity of $7 \cdot 10^5$ Sm/m.

From the figure, it can be seen that the calculated and experimental data diverge. First of all, this is observed for the resonant length of the half-wave wire. For the microwire, these values are shifted toward the range 0.4–0.42 λ, which testifies to the delay of the wave propagating along the wire. In [33] the assumption was put forward that this is caused by the influence of the nanostructured transitive layer. At present theoretical estimation of the influence of this layer is still impossible despite the successful development of such direction, as nanoelectrodynamics [34]. Therefore, at present the main method for investigation of such objects is experimental, and perhaps, it is difficult to find for then method of research other than with the use of the OR.

## 7.5 Study of Small Spherical Objects

One of the most widespread shapes of inclusions during synthesis of composite materials is a sphere. It is of interest to measure the polarizability of such object at a preset frequency with the use of the open resonator.

To elucidate the influence of a small sphere on the open resonator characteristics, we take advantage of the test-object method suggested in [25] for measuring the field distribution in the space between the mirrors. If the sphere with small electric diameter is placed at the point at which the electric and magnetic field strengths of the own oscillations are equal to $E$ and $H$, then in accordance with [25], the resonant oscillation frequency of the open resonator can be written in the form:

$$\Delta f = -\frac{f}{2N}\left(\varepsilon_0 \alpha_e E^2 - \mu_0 \alpha_m H^2\right). \tag{7.16}$$

Here $\alpha_e$ and $\alpha_m$ are the electric and magnetic polarizabilities of the sphere. For spheres with small intrinsic losses, these quantities can be considered real.

If the sphere is placed at the antinode of the electric or magnetic oscillation field of the open resonator of the given type and the corresponding resonant frequency shift is measured, then the electric and magnetic polarizabilities of the examined samples can be calculated from the formulas:

$$\alpha_e = -\frac{\pi w_0^2 L}{2}\frac{\Delta f'_e}{f}, \tag{7.17}$$

$$\alpha_m = -\frac{\pi w_0^2 L}{2}\frac{\Delta f'_m}{f}. \tag{7.18}$$

Since the method is approximate, it is expedient to verify it by solving the direct problem. To this end, the resonant frequency shifts were measured in the OR in the

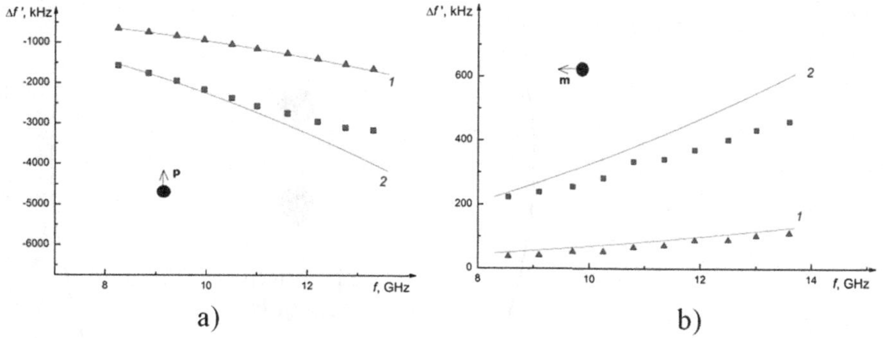


**Fig. 7.9** Resonant frequency shift of the open resonator with copper spheres. The solid curves show the calculated results, and symbols are for measurements with sphere diameters of (**a**) 4.5 mm (curve *1*) and 6.0 mm (curve *2*) and (**b**) 2.6 mm (curve *1*) and 4.0 mm (curve *2*). Here **p** and **m** denote the electric and magnetic dipole moments

range 8–15 GHz for copper spheres, and the results obtained were compared with the data of calculations using Eq. (7.16) (Fig. 7.9).

As can be seen from the figure, good agreement between the measured and calculated data is observed for the sphere 4.5 mm in diameter in the entire frequency range, and for the sphere 6 mm in diameter, the curves start to diverge in the high frequency range. This allows us to determine the allowable frequency range for measurements of the parameters of spherical samples, since the metal sphere introduces the greatest perturbations into the open resonator in comparison with nonmetallic materials. In this case, in the electric field, it was possible to determine the real polarizability component for spheres 6 mm in diameter at frequencies 11–12 GHz, whereas in the magnetic field, it was possible to measure correctly only the parameters of spherical samples 4 mm in diameter at frequencies up to 9 GHz. These relationships between the particle sizes and the employed frequency range can be generalized toward shorter wavelengths of the electromagnetic spectrum.

The same method was used to measure the polarizabilities of spherical aerogel samples made of multiwall carbon nanotubes [35]. As demonstrated measurements, values of the electric and magnetic polarizabilities of spherical aerogel samples were less than those of metal spheres of the same diameter by about 10%. These results have been obtained for the first time.

In the open resonator, it is possible to measure polarizabilities of not only single spheres but also clusters of two and more objects (Fig. 7.10).

When the spheres are symmetric about the resonator axis, they appear at the points with the same fields. In this case, the polarizability becomes tensor in character; therefore, it is necessary to consider the cluster arrangement with respect to the direction of the electric and magnetic field strength vectors:

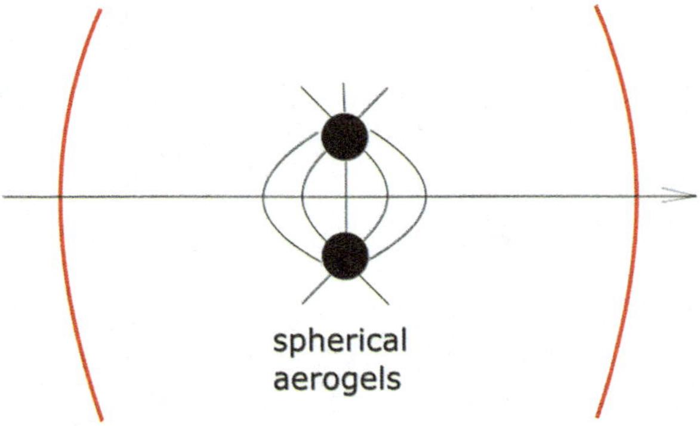


**Fig. 7.10** Open resonator with two spherical objects

$$\alpha_{e,cl} = 2\alpha_e = \alpha^e_{xx} = \alpha^e_{yy} = -\frac{\pi w_0^2 L}{2} \frac{\Delta f'_{e,cl}}{f\exp(-2l^2/w_0^2)}, \tag{7.19}$$

$$\alpha_{m,cl} = 2\alpha_m = \alpha^m_{xx} = \alpha^m_{yy} = -\frac{\pi w_0^2 L}{2} \frac{\Delta f'_{m,cl}}{f\exp(-2l^2/w_0^2)}. \tag{7.20}$$

Here $l$ is the distance from the OR axis to one of the spheres.

In [36], the interaction of spherical aerogel samples was studied by this method, and the electric polarizability components of the cluster of two samples were measured. As a result, measurements have demonstrated that when the electric field vector is parallel to the axis connecting the aerogel sample centers, the components of the electric polarizability tensor are equal to the sum of polarizabilities of individual samples up to the minimal distances between them. If the electric field vector is perpendicular to the axis connecting the sphere centers, then for distances between the spheres less than about half wavelength, the tensor component differed from the sum of the polarizabilities of the objects. The measured polarizabilities can be used to produce artificial materials with inclusions of spherical aerogel samples made of carbon nanotubes.

## 7.6 Inclusion of a Thin Film in the OR

Insertion of the thin dielectric film with thickness $d$ and dielectric permittivity $\varepsilon$ into the space between the mirrors leads to the resonant frequency shift and decrease of the Q-factor. The latter seems to be very insignificant if the film is thin and uniform and is made of a dielectric with small losses. However, in the experiment with the film strictly parallel to the plane of mirrors, significant effect of the film position on the longitudinal axis (or its movement along the axis) of the resonator was sometimes observed, and the Q-factor decreased noticeably for some positions of the film.

This is explained by the fact that when the film is parallel to the mirrors, three coupled resonators are formed rather than one (two resonators are formed by each of the mirrors and the film, and the third resonator is formed, as previously, by the two mirrors). At each coincidence of the frequencies of these resonators, we observe the mode of coupled oscillations, well-known in radio engineering and accompanied by resonant curve broadening. Therefore, the dielectric film tilted at the angle $\vartheta$ for which no coupled resonators arise will be more convenient for investigations, measurements, and continuous control.

However, for a tilted film, a portion of the wave reflected by it is incident at the mirrors and, due to diffraction at the mirror ends, can return to the resonant oscillation; in this case, when returning, it can be both in phase and antiphase. The power scattered by the film can be excluded completely by using sufficiently large tilt angles.

For different polarization directions of the wave excited in the resonator, the conditions of passage through the tilted film will differ; correspondingly, the resonant frequency shifts and changes of the Q-factor caused by the film will also differ. Thus, for the polarization parallel to the film plane, these parameters are described by the formulas:

$$\delta f_{II} = f_0 \frac{d}{L_{\mathrm{res}}} \frac{\varepsilon - 1}{\varepsilon \cos \vartheta}, \tag{7.21}$$

$$2\Delta f_{II} = \frac{\pi f_0^2 d^2}{2cL_{\mathrm{res}}} \frac{(\varepsilon - 1)^2}{\cos^2 \theta}. \tag{7.22}$$

For the polarization of the electric vector in the plane perpendicular to the film plane, it is easy to obtain:

$$\delta f_{\perp} = f_0 \frac{d}{L_{\mathrm{res}}} \frac{(\varepsilon - 1)[(\varepsilon + 1)\cos^2\vartheta + 1]}{2\varepsilon \cos \vartheta}, \tag{7.23}$$

$$2\Delta f_{\perp} = \frac{\pi f_0^2 d^2}{2cL_{\mathrm{res}}} \frac{(\varepsilon - 1)^2[(\varepsilon + 1)\cos^2\vartheta + 1]^2}{\varepsilon^2 \cos^2\vartheta}. \tag{7.24}$$

All these formulas have been derived under assumption that the tilted film provides complete extraction of the radiation power reflected by the film from the resonator. If the angle $\vartheta$ is small and the wave reflected by the film remains in the resonator, the interference phenomena we discussed above are inevitable. This is well illustrated by a comparison of the calculated and experimental data [37] shown in Figs. 7.11 and 7.12, where the dashed curves show the resonant frequency shifts and losses calculated from Eqs. (7.21), (7.22), (7.23), and (7.24) and the solid curves show the measured values obtained when the film moved along the longitudinal resonator axis.

It can be seen that at tilt angles in the range from 0° to 30°, the backward wave reflected from the film causes interference and dependence of the resonator parameters on the film position (vibration) along the longitudinal OR axis

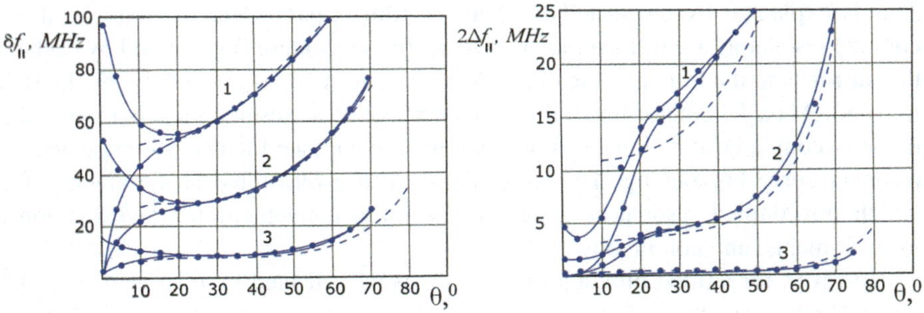


a) b)

**Fig. 7.11** Dependence of the detuning (**a**) and losses (**b**) on the tilt angle of the film that moved along the resonator axis (for the parallel polarization): 1 and 2 are for$\varepsilon = 2.66$ and $d = 300$ and 150 μm, and 3 is for $\varepsilon = 2.25$ and $d = 50$ μm

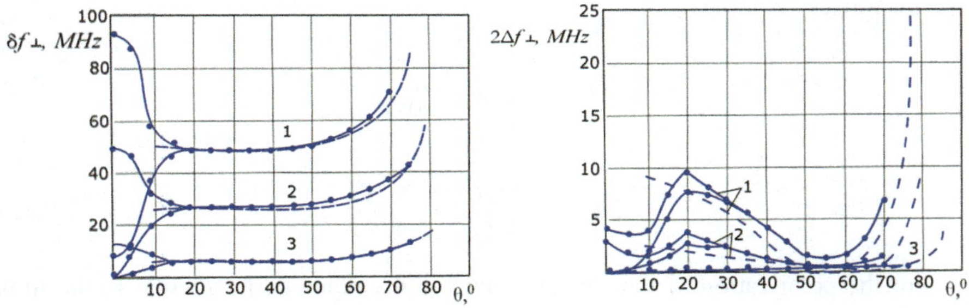


a) b)

**Fig. 7.12** Dependence of the detuning (**a**) and losses (**b**) on the tilt angle of the film that moved along the resonator axis (for the orthogonal polarization): 1 and 2 are for $\varepsilon = 2.66$ and $d = 300$ and 150 μm, and 3 is for $\varepsilon = 2.25$ and $d = 50$ μm

(hatched regions). For large $\vartheta$ and polarization of the $E$ field in the plane orthogonal to the film surface (Fig. 7.12a), the measured resonator frequency shift is indifferent to the film position or vibrations and depends only on the film thickness and the $\varepsilon$ value. Moreover, it turns out that for the orthogonal polarization, the resonant frequency shift is also indifferent to angular vibrations for sufficiently large range of tilt angles. This mode is very convenient for measurements of the dielectric film permittivity (or film thickness for the known $\varepsilon$ value) as well as for control over the dynamics of its parameters when the film is stretched between the mirrors and thus naturally vibrates.

## 7.7 Investigation of a Composite Material

Arrangement of a composite sample in space between the mirrors is expedient only when the losses introduced by the sample after a single pass of the wave will be insignificant. These situations can arise if the sample is small and can be approximated by a small sphere or by a small elongated scatterer; we have already considered these cases above. The composite can also represent a sufficiently transparent thin layer analogous to that we have already considered as well. If the composite, on the contrary, is not transparent and possesses so high conductivity that its surface represents a very good reflector, it can be used as one of the resonator mirrors (Fig. 7.13). In all these cases, the arrangement of the examined sample into the

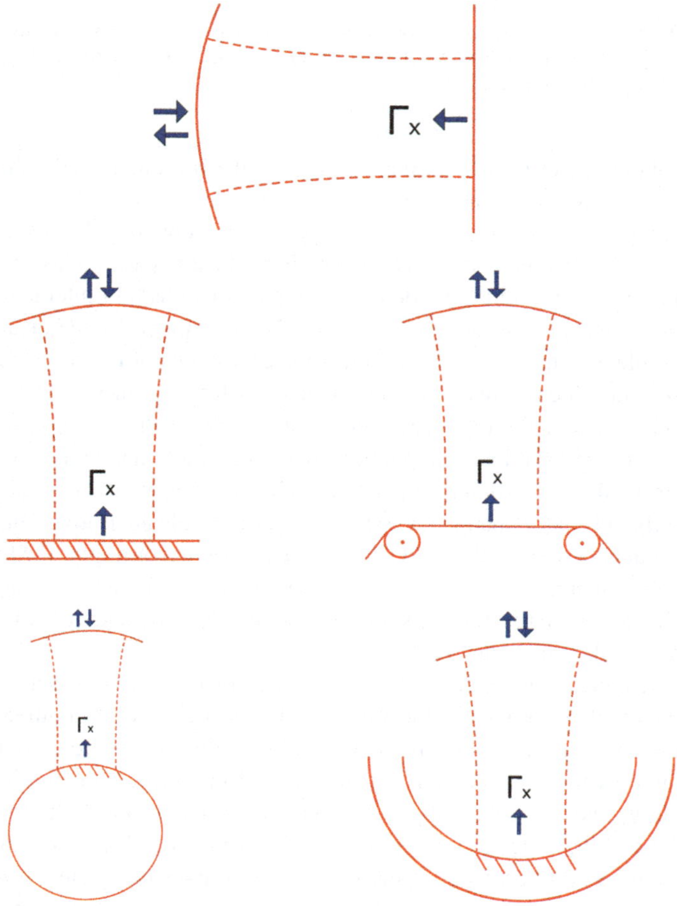


**Fig. 7.13** Examined objects of different configurations used as one of the mirrors of the open resonator

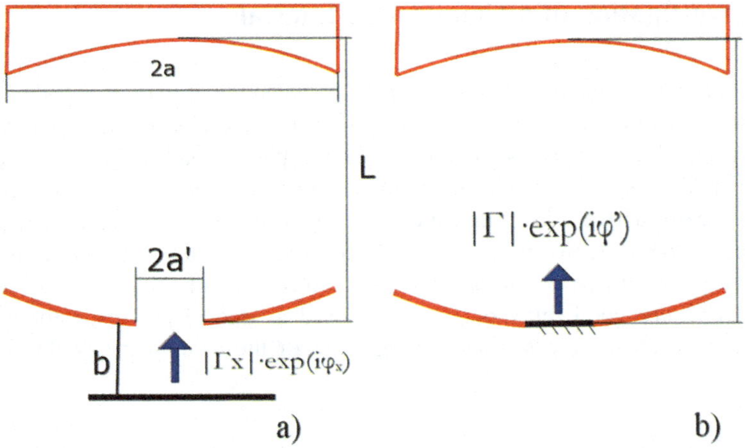


**Fig. 7.14** Open resonator with a hole in one of the mirrors for external measurements: (**a**) scheme of examined object arrangement under the measuring hole and (**b**) mathematical model of the resonator with a combined mirror

resonator retains the Q-factor and the structure of the resonator field, which makes the application of the resonant technique efficient.

If the sample of a composite material possesses considerable absorption, the application of the resonator becomes inexpedient. In this case, measurements can be performed in a beam in the mode of single pass or reflection interaction. In this case, the local character of measurements will be determined by the beam aperture which, as a rule, changes from several units to several tens of the wavelengths. It is possible to obtain local reflection coefficient for a large surface of the composite material sample using the open resonator with a hole in the center of one of the mirrors. Placing the examined sample behind this hole, we change the conditions of reflection from this part of the mirror; naturally, the resonant frequency and the Q-factor of the resonator change correspondingly. The hole aperture is chosen much smaller than the aperture of the field spot on the mirror; accordingly, the Q-factor of the resonator remained sufficiently high even if the examined material strongly absorbed the wave energy during single interaction. The scheme of this measurement method is shown in Fig. 7.14a.

The mathematical OR model with the examined material located behind the hole can be developed by replacing the mirror with the hole by the equivalent combined mirror whose reflection coefficient is variable, Fig. 7.14b. Beyond the hole, it is expedient to assume that the mirror is ideally reflecting (the reflection modulus is equal to unity, and the phase angle is equal to $\pi$), and in the hole aperture, the modulus $|\Gamma'|$ and the reflection phase $\varphi'$ can take arbitrary values depending on the properties of the examined material and its spacing from the hole plane. A solution of the integral equation for such resonator [38] allows us to write the expression for losses in the resonator caused by such equivalent mirror:

$$\eta_x = 1 - \left[c^2 + |\Gamma'|^2(1-c)^2 + 2|\Gamma'|c(1-c)\cos\varphi'\right], \tag{7.25}$$

where $c = \exp\left[-(a')^2/w_m^2\right]$, $a'$ is the hole radius, and $w_m$ is the radius of the field spot on the mirror with hole. The uncertainty due to the phase angle $\varphi'$ can be eliminated by moving the sample behind the hole along the longitudinal resonator axis and fixing maximal and minimal values of its Q-factor. The maxima of the Q-factor correspond to the minimal losses at $\varphi = 2\pi n$ ($n = 0, 1, 2, 3\ldots$):

$$\eta_{\min} = 1 - [c + |\Gamma'|(1-c)]^2, \tag{7.26}$$

and the minima of the Q-factor correspond to the maximal losses:

$$\eta_{\max} = 1 - [c - |\Gamma'|(1-c)]^2. \tag{7.27}$$

From these formulas, the required values of $|\Gamma'|$ can easily be calculated. To exclude the losses in the hole itself and to determine the coefficient of reflection from the region of the examined material located behind the hole, it is expedient first to place a metal sheet rather than the examined sample behind the hole, taking for it $|\Gamma| = 1$ and $\varphi = \pi$, and to perform the corresponding calibration.

Thus, the procedure of investigation of the composite material sample (representing a plane plate of fixed thickness or, if the sample has some other shape, it should have a plane cut – a plane side along the surface of which measurements are performed) is reduced to measurements at first the maximal and minimal Q-factors (the transfer coefficient) of the resonator when moving the plane metal sheet behind the hole with simultaneous fixation of distances between the sheet plane and the rare mirror side corresponding to these maxima and minima. Then the examined sample of the composite material is arranged behind the hole, and analogous measurements are performed for each local sample fragment. In such a way, the distribution of the modulus and phase of the reflection coefficient over the sample at a preset frequency can be investigated. Despite the seeming complexity of the given technique, it is easily automated and, which is the most important, allows one to obtain very useful information on the examined sample. As an example, Fig. 7.15 shows the distribution of these parameters obtained for the experimental sample made of the composite Plexiglas material filled with threadlike carbon segments.

These results testify to a wide spread of values of the reflection coefficient modulus over the sample as well as to a significant phase inhomogeneity of this sample. It is obvious that if one tries to measure its parameters in a large aperture beam using the standard free space technique integrating the signal reflected from such sample, the error will be very large because the phase portrait of this plane sample is analogous to a lens.

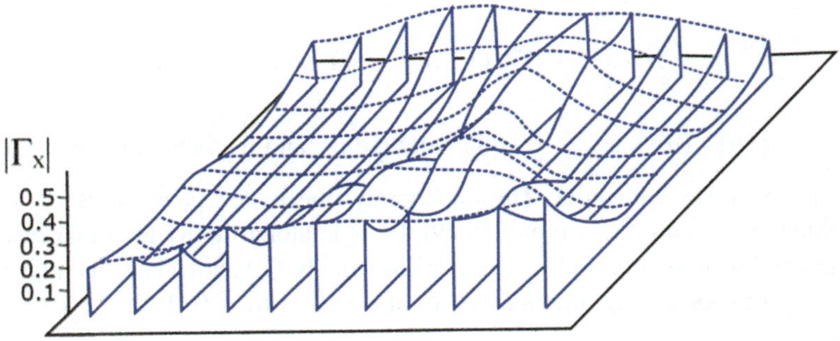


**Fig. 7.15** Distribution of the modulus of the reflection coefficient $|\Gamma_x|$ over the surface of the plane composite measured using the OR with a hole

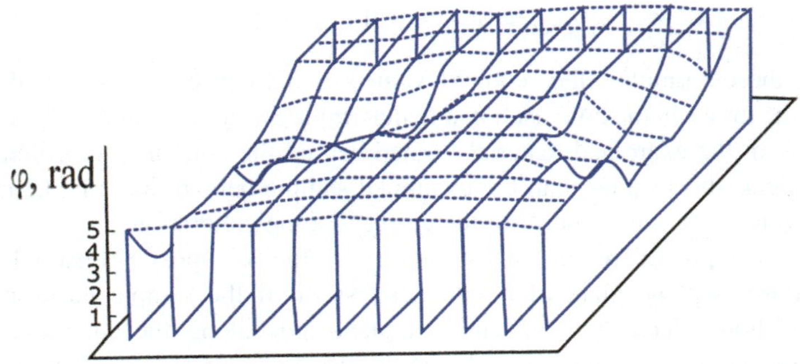


**Fig. 7.16** Distribution of the phase of the reflection coefficient $\varphi_x$ over the surface of the plane composite measured using the OR with a hole

## 7.8 Sensitivity of the Method: Optimization of Resonator Tuning

Measurements in the open resonator are attractive due to their high sensitivity, simplicity, and convenience of sample change. However, the measurement conditions call for a serious analysis. On the one hand, the OR ensures great additional capabilities of investigation of new materials and their components, but on the other hand, as we will see below, these capabilities are realized not for all modes of resonator tuning. Analysis of the method sensitivity is especially important when it is used to investigate very small objects introducing small perturbations into the resonator field or to investigate small oscillations of the object parameters with respect to their steady state. Anyway, we assume that we need to measure in statics or to trace in dynamics very small changes of the resonator Q-factor and (or) its resonant frequency.

In spite of the fact that the OR is multifrequency, we will consider the behavior of only one oscillation mode considering that its resonant curve is sufficiently distinguished against the background of the neighboring modes; this problem is simply solved technically. We consider that the initial resonant frequency of this oscillation $f_0$ is preset and that its Q-factor is known: $Q_0 = 2k_1 L_{res}/\eta_0$. We note that the losses for circular pass of the resonator include all types of losses except those on coupling elements. If we want to estimate small changes of losses caused by the object placed into the resonator, the stationary losses introduced by the given object are assumed to be taken into account in $\eta_0$. We designate by $\Delta\eta_x$ small changes of losses introduced by the given object that we want to detect. When the examined object is small, we also describe by $\Delta\eta_x$ the losses introduced by the small object.

Changes occurring in the resonator can be estimated either from the change of its transmission coefficient that calls for the presence of two coupling elements – input and output ones – or from the change of the reflection coefficient; in this case, it is sufficient to use only one coupling element of the resonator with the external transmission line. One of the special features of the open resonator is that the coupling elements in the form of holes or slits in the resonator mirrors can introduce diffraction losses in addition to the coupling losses of the resonator with an external transmission line. These additional losses depend on the configuration of the coupling element; it is possible to consider their value approximately proportional to the useful coupling losses. As a result, we will describe by $\eta_{n_1}(1+\varsigma_1)$ and $\eta_{n_2}(1+\varsigma_2)$ the losses introduced by each coupling element, where $\varsigma$ characterizes the fraction of the diffraction losses on the coupling elements. It is possible to avoid these diffraction losses using translucent mirrors or exciting the resonator using a tilted dielectric film, but in these cases, additional losses will be due to scattering on the mirrors and film. However, in this case the coefficients of coupling of the resonator with the transmission line will be determined only by the radiation losses in the line on the input and output coupling elements: $\beta_1 = \eta_{n_1}/\eta_0$ and $\beta_2 = \eta_{n_2}/\eta_0$. As a result, the power transmission and reflection coefficients can be written in the form:

$$|T|^2 = \frac{4\beta_1\beta_2}{[2+\beta_1(1+\varsigma_1)+\beta_2(1+\varsigma_2)]^2 + 4Q_0^2\delta_f^2}, \tag{7.28}$$

$$|\Gamma|^2 = 1 - \frac{4\beta_1[1+\varsigma_1\beta_1+(1+\varsigma_2)\beta_2]}{[1+\beta_1(1+\varsigma_1)+\beta_2(1+\varsigma_2)]^2 + 4Q_0^2\delta_f^2}, \tag{7.29}$$

where

$$\delta_f = \frac{f_{osc} - f_0}{f_0} \tag{7.30}$$

describes the relative detuning of the external source (generator) from the resonant frequency of the resonator.

Let us assume that the losses in the resonator change by a small value $\Delta\eta_x$. In this case, we expect changes of the resonator transmission and reflection coefficients. We describe this process by the parameters $S_{T\eta}$ and $S_{\Gamma\eta}$ that characterize the sensitivity of

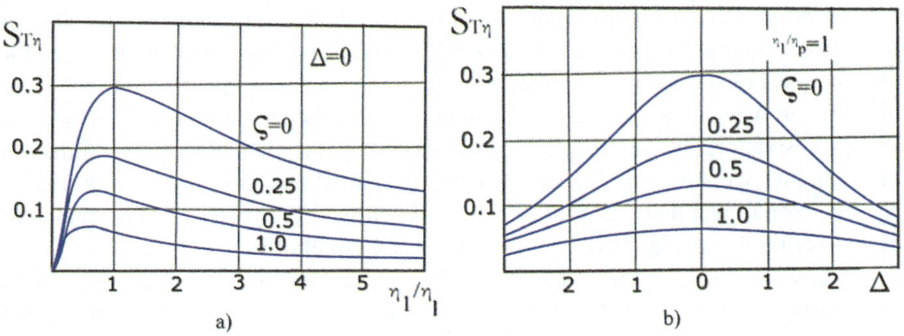


**Fig. 7.17** Sensitivity of diagnostics of small variations in the resonator losses $\eta$ from the change of the transmission coefficient $T$: (**a**) at the indicated values of the coupling coefficient $\beta_1 = \eta_1/\eta_0$ and (**b**) at the indicated values of the relative resonator detuning $\Delta = 2Q\delta_f$

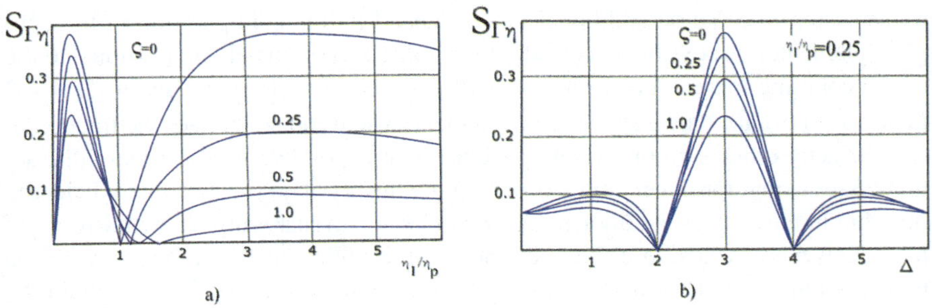


**Fig. 7.18** Sensitivity of the diagnostics of small variations in the resonator losses $\eta$ to the change of the reflection coefficient $\Gamma$: (**a**) at the indicated values of the coupling coefficient $\beta_1 = \eta_1/\eta_0$ and (**b**) at the indicated values of the relative resonator detuning $\Delta = 2Q\delta_f$

resonator transmission and reflection characteristics to small changes of the introduced losses: $\Delta|T|^2 = S_{T\eta} \cdot \Delta\eta_x$ and $\Delta|\Gamma|^2 = S_{\Gamma\eta} \cdot \Delta\eta_x$.

The dependences of the sensitivity parameters $S_{T\eta}$ and $S_{\Gamma\eta}$ on the relative detuning $\Delta^2 = 4Q_0^2\delta_f^2$ and on the coupling coefficient at the resonator input are shown in Figs. 7.17 and 7.18.

The dependence of the sensitivity to small variations in the resonator losses for the transmission coefficient $S_{\Gamma\eta}$ has a maximum at zero detuning ($\Delta = 2Q\delta_f = 0$) and $\beta_1 = \beta_2 = 1/(1 + \varsigma)$.

For the scheme with reflection, the dependence of the sensitivity to small variations in resonator losses is shown in Fig. 7.18. Here zero sensitivities are observed at tuning in resonance (with coupling close to critical one $\beta_1 \approx 1$) and at detuning on the slopes of the resonant curve. The sharp sensitivity maximum can be observed at zero detuning for the coupling coefficient $\beta_1 = \left(2 \pm \sqrt{3 + \varsigma_1^2}\right)/(1 - \varsigma_1)$.

The above formulas allow the sensitivity to small changes of the phase shift introduced by the examined object to be determined. Examples of dependences of

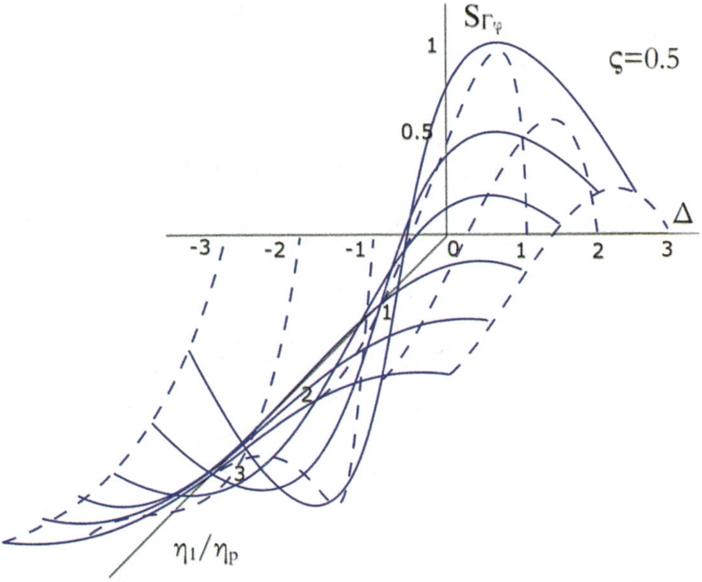


**Fig. 7.19** Sensitivity of the resonator reflection coefficient to variations in the phase shift introduced by the controllable object depending on the resonator coupling coefficient $\beta_1 = \eta_1/\eta_0$ and on the relative detuning $\Delta = 2Q\delta_f$

the sensitivity of the transmission coefficient $S_{T\varphi}$ and reflection coefficient $S_{\Gamma\varphi}$ to small variations of the phase incursion in the resonator are shown in Figs. 7.19 and 7.20.

It can be seen that the maximum sensitivity in both cases is observed on the slopes of the resonant curve rather than in resonance, and the sensitivity for the scheme with reflection can be even higher. The optimal values of the coupling coefficient for the scheme with transmission are $\beta_1 = \beta_2 = 1$ and, for the scheme with reflection, $\beta_1 = \left(\sqrt{3}-1\right)/\left\lfloor 4 - \varsigma\left(\sqrt{3}-1\right)\right\rfloor$.

Thus, we see that to obtain high sensitivity to variations of different (in amplitude or phase) changes when investigating small objects or small variations of the parameters introduced in the OR by objects, it is necessary to provide optimal conditions of coupling and tuning of the open resonator.

## 7.9 Conclusions

From the examples considered above, it follows that taking advantage of multiple interaction of the electromagnetic wave with the examined object, it is really possible to obtain useful information on the parameters of small components of composite materials and on the parameters of the composite as a whole. This technique is especially attractive due to the fact that unlike the techniques using volume

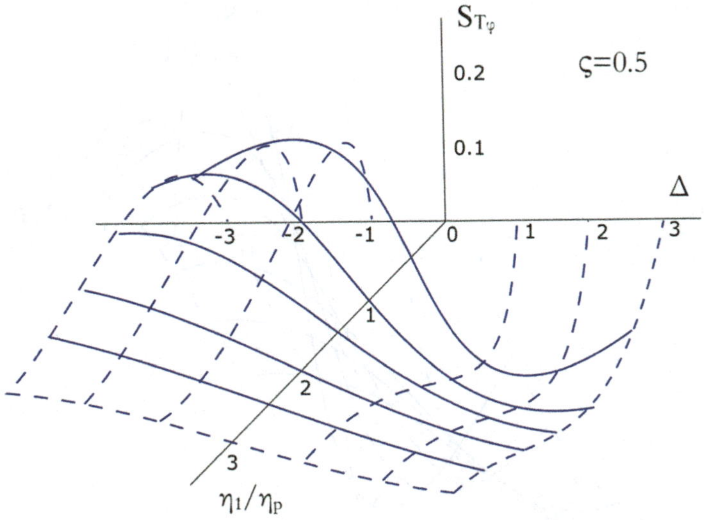


**Fig. 7.20** Sensitivity of the resonator transmission coefficient to variations in the phase shift introduced by the controllable object depending on the resonator coupling coefficient $\beta_1 = \eta_1/\eta_0$ and on the relative detuning $\Delta = 2Q\delta_f$

resonators, it is very convenient in laboratory studies, since it allows the examined samples to be easily changed in the working space of the open resonator. This technique is especially promising for industrial applications, since it allows nondestructive contactless control of small material samples, different extended superthin threadlike or film components, and finished composite structures to be performed. Control can be carried out in dynamics, in the course of their production.

Despite significant advantages of the quasi-optical resonator technique, the number of authors that implement it is not large. This demonstrates that the application of the resonator methods in the terahertz frequency range faces experimental difficulties; in particular, it requires very stable radiation sources. The application of vector network analyzers (VNA) is limited by the low-frequency part of the terahertz range [39, 40, 10], since the effect of repeated frequency multiplication is employed and the output signal power of such devices is very low. Sources based on backward wave oscillators (BWO) have higher output power, and the scheme of frequency stabilization ensures resolution comparable with that of the VNA [11]. However, the practical application of this equipment is limited by a frequency of ~1.5 THz. At present the time-domain spectroscopy with pulsed sources of electromagnetic radiation at higher frequencies is used [41]. However, incorporation of the open resonator into the path of the time-domain spectrometer faces significant difficulties and, as far as we know, has not realized yet.

In our opinion, with further mastering of millimeter and submillimeter (terahertz) ranges, occurrence in the market of inexpensive generators, detectors, and other element base for these ranges, the quasi-optical resonator methods of measurements, diagnostics, and nondestructive contactless control will become more popular in research and industry.

## References


1. Dicke RH (1958) Molecular amplification and generation systems and methods. USA
2. Yang X, Kweun JM, Kim YY (2018) Theory for perfect transmodal Fabry-Perot interferometer. Sci Rep 8(1):69. https://doi.org/10.1038/s41598-017-18408-5
3. Liu D-Y, Li M-H, Zhai X-M, Yao L-F, Dong J-F (2014) Enhanced asymmetric transmission due to Fabry-Perot-like cavity. Opt Express 22(10):11707–11712 Retrieved from http://www.ncbi.nlm.nih.gov/pubmed/24921293
4. Kweun JM, Lee HJ, Oh JH, Seung HM, Kim YY (2017) Transmodal Fabry-Pérot resonance: theory and realization with elastic metamaterials. Phys Rev Lett 118(20):205901. https://doi.org/10.1103/PhysRevLett.118.205901
5. Xiao Z, Liu D, Ma X, Wang Z (2015) Multi-band transmissions of chiral metamaterials based on Fabry-Perot like resonators. Opt Express 23(6):7053–7061 Retrieved from http://www.ncbi.nlm.nih.gov/pubmed/25837050
6. Liang W, Bockrath M, Bozovic D, Hafner JH, Tinkham M, Park H (2001) Fabry – Perot interference in a nanotube electron waveguide. Nature 411(6838):665–669. https://doi.org/10.1038/35079517
7. Li C, Lan T, Yu X, Bo N, Dong J, Fan S (2017) Room-temperature pressure-induced optically-actuated Fabry-Perot nanomechanical resonator with multilayer graphene diaphragm in air. Nanomaterials 7(11):366. https://doi.org/10.3390/nano7110366
8. Culshaw W (1962) Measurement of permittivity and dielectric loss with millimetre-wave Fabry-Perot interferometer. Proc IEE – B Electron Commun Eng 109(23):6. https://doi.org/10.1049/pi-b-2.1962.0140
9. Dhillon SS, Vitiello MS, Linfield EH, Davies AG, Hoffmann MC, Booske J et al (2017) The 2017 terahertz science and technology roadmap. J Phys D Appl Phys 50(4):43001. https://doi.org/10.1088/1361-6463/50/4/043001
10. Sun W, Yang B, Wang X, Zhang Y, Donnan R (2013) Accurate determination of terahertz optical constants by vector network analyzer of Fabry–Perot response. Opt Lett 38(24):5438. https://doi.org/10.1364/OL.38.005438
11. Gao YC, Jones CR, Dutta JM (2009) Phase locked BWO system for open resonator measurement in D-Band. IEEE Microw Wirel Comp Lett 19(9):599–601. https://doi.org/10.1109/lmwc.2009.2027098
12. *IEEE P1785.2/D1n, April 2015 : IEEE draft standard for rectangular metallic waveguides and their interfaces for frequencies of 110 GHz and above – part 2: waveguide interfaces* (2015) IEEE. Retrieved from http://ieeexplore.ieee.org/document/7109075/
13. *IEEE Std 1785.3-2016 : IEEE recommended practice for rectangular metallic waveguides and their interfaces for frequencies of 110 GHz and above – part 3: recommendations for performance and uncertainty specifications* (2016) IEEE. Retrieved from http://ieeexplore.ieee.org/document/7589955/
14. Kotelnikov I, Altynnikov A, Mikhaylov A, Medvedeva V, Kozyrev A (2016) Electrodeless measurement technique of complex dielectric permittivity of high-K dielectric films in the millimeter wavelength range. Prog Electromagn Res M 52:161–167. https://doi.org/10.2528/pierm16100505
15. Kogelnik H, Li T (1966) Laser beams and resonators. Proc IEEE 54(10):1312–1329. https://doi.org/10.1109/PROC.1966.5119

16. Shalygina EE, Kharlamova AM, Shalygin AN, Molokanov VV, Umnov PP, Umnova NV, Chueva TR (2016) Structural and magnetic properties of "thick" microwires produced by the modernized Ulitovsky–tailor method. J Magn Magn Mater 415:106–110. https://doi.org/10.1016/J.JMMM.2016.01.056
17. Kraus L, Frait Z, Ababei G, Chayka O, Chiriac H (2012) Ferromagnetic resonance in submicron amorphous wires. J Appl Phys 111(5):53924. https://doi.org/10.1063/1.3693031
18. Ramayya EB, Vasileska D, Goodnick SM, Knezevic I (2007) Electron mobility in silicon nanowires. IEEE Trans Nanotechnol 6(1):113–117. https://doi.org/10.1109/TNANO.2006.888521
19. Zhukov A, Shuvaeva E, Kaloshkin S, Churyukanova M, Kostitsyna E, Zhdanova M et al (2016) Studies of interfacial layer and its effect on magnetic properties of glass-coated microwires. J Electron Mater 45(5):2381–2387. https://doi.org/10.1007/s11664-015-4319-y
20. Huang Q, Lilley CM, Bode M, Divan RS (2008) Electrical properties of Cu nanowires. In: 2008 8th IEEE conference on nanotechnology. IEEE, pp 549–552. https://doi.org/10.1109/NANO.2008.163
21. Steinhoegl W, Schindler G, Steinlesberger G, Traving M, Engelhardt M (2003). Scaling laws for the resistivity increase of sub-100 nm interconnects. In: International conference on simulation of semiwire processes and devices. *SISPAD 2003,* IEEE, pp 27–30. https://doi.org/10.1109/SISPAD.2003.1233629
22. Sondheimer EH (1952) The mean free path of electrons in metals. Adv Phys 1(1):1–42. https://doi.org/10.1080/00018735200101151
23. Mayadas AF, Shatzkes M, Janak JF (1969) Electrical resistivity model for polycrystalline films: the case of specular reflection at external surfaces. Appl Phys Lett 14(11):345–347. https://doi.org/10.1063/1.1652680
24. Marin P, Cortina D, Hernando A (2008) Electromagnetic wave absorbing material based on magnetic microwires. IEEE Trans Magn 44(11):3934–3937. https://doi.org/10.1109/tmag.2008.2002472
25. Weinstein LA (1969) Open resonators and open waveguides. Golem Press, Boulder
26. Dorofeev IO, Dunaevskii GE (2012) Losses in an open microwave resonator with an ultrathin cylinder. Russ Phys J 54(10):1112–1120
27. Vázquez M, Adenot-Engelvin A-L (2009) Glass-coated amorphous ferromagnetic microwires at microwave frequencies. J Magn Magn Mater 321(14):2066–2073. https://doi.org/10.1016/J.JMMM.2008.10.040
28. Dorofeev IO, Dunaevskii GE, Lebedev IA, Polozov GG (2012) Open Quasioptic microwave resonator with a cast glass-coated magnetic microwire. Russ Phys J 55(9/2):30–33 (In Russian)
29. Dorofeev IO, Dunaevskii GE (2014) A device for the quasi-optical resonance diagnostics of a glass-coated cast microwire. Russ J Nondestruct Test 50(12):741–747. https://doi.org/10.1134/s1061830914120031
30. Dorofeev IO, Dunaevskii GE, Larin VS (2015) Scattering of the field of an open quasi-optical microwave resonator on microwire with inhomogeneities. Russ Phys J 58(8):1167–1171. https://doi.org/10.1007/s11182-015-0627-1
31. Lopez-Dominguez V, Garcia MA, Marin P, Hernando A (2017) Tuning metamaterials by using amorphous magnetic microwires. Sci Rep 7:9. https://doi.org/10.1038/s41598-017-09665-5
32. Nesterenko MV (2010) Analytical methods in the theory of thin impedance vibrators. Prog Electromag Res B 21:299–328
33. Dorofeev IO, Dunaevskii GE (2017) Resonance characteristics for microwire pieces as elements of composite materials. Russ Phys J 59(12):2080–2086. https://doi.org/10.1007/s11182-017-1017-7
34. Slepyan GY, Maksimenko SA, Lakhtakia A, Yevtushenko O, Gusakov AV (1999) Electrodynamics of carbon nanotubes: dynamic conductivity, impedance boundary conditions, and surface wave propagation. Phys Rev B 60(24):17136–17149. https://doi.org/10.1103/PhysRevB.60.17136

35. Kuznetsov VL, Suslyaev VI, Dorofeev IO, Kazakova MA, Moseenkov SI, Smirnova TE, Krasnikov DV (2015) Investigation of electromagnetic properties of MWCNT aerogels produced via catalytic ethylene decomposition. Phys Status Solidi B-Basic Solid State Phys 252 (11):2519–2523. https://doi.org/10.1002/pssb.201552254
36. Krasnikov DV, Dorofeev IO, Smirnova TE, Suslyaev VI, Kazakova MA, Moseenkov SI, Kuznetsov VL (2017) Electromagnetic interaction between spherical aerogels of multi-walled carbon nanotubes. Phys Status Solidi B:1700256. https://doi.org/10.1002/pssb.201700256
37. Dunaevskii GE (2006) An open Quasioptical transformers. Izdatelstvo NTL, Tomsk (in Russian)
38. Valitov RA, Dubko SF, Kamyshan VV, Kuzmichev VM, Makarenko VI, Sokolov AV, Sheyko VP (1969) In: Valitov RA (ed) Submillimeter wave technology. Sovetskoe Radio, Moscow (in Russian)
39. Yang BB, Katz SL, Willis KJ, Weber MJ, Knezevic I, Hagness SC, Booske JH (2012) A high-Q terahertz resonator for the measurement of electronic properties of wires and low-loss dielectrics. IEEE Trans Tera Sci Tech 2(4):449–459. https://doi.org/10.1109/TTHZ.2012.2199578
40. Suzuki H, Kamijo T (2008) Millimeter-wave measurement of complex permittivity by perturbation method using open resonator. IEEE Trans Instrum Meas 57(12):2868–2873. https://doi.org/10.1109/tim.2008.926448
41. Zhang X-C, Xu J (2010) Introduction to THz wave photonics. Springer US, Boston. https://doi.org/10.1007/978-1-4419-0978-7

# Chapter 8
# Wave Vision

**Vladimir P. Yakubov**

**Abstract** The basic applications of radio waves are communication and radiolocation. In the proposed chapter, another application of radio waves is considered – radio wave tomography – which is based on penetration of radio waves into opaque media and the extraction of information about the three-dimensional internal structure of objects. The theory of extracting useful information in this case is solving inverse problems using aperture synthesis with focusing. The efficiency of this approach is illustrated by the numerous examples.

## 8.1 Review of Existing Approaches

In this chapter we present development of wave tomography methods as a means of remote nondestructive testing, diagnostics of the internal structure of semitransparent media, and reconstruction of the shapes of opaque objects based on methods of radiolocation. The word tomography derives from two Greek words: τομος, a layer, and γραφειν, to write. Thus, tomography means literally "to write a layer," i.e., to investigate a structure layer-by-layer. The difference between tomography and other diagnostic methods is that information from the same test element is recorded in multiple integral projections, i.e., many times from different angles relative to the embedded inhomogeneities.

It has been over 50 years now since researchers in this field learned how to "clear up" these projections and recover the structure of inhomogeneities layer-by-layer. For the most part, this became possible, thanks to the development of new computational methods and computer technologies. At the present time, computed tomography is rightfully considered as an "absolute" diagnostic technique in medicine. Radio wave tomography is similar to X-ray and magnetic resonance imaging, but it is associated with electromagnetic radiation in the radio wave range. In this case, the wavelength of the radiation used is comparable to the characteristic size of the

V. P. Yakubov (✉)
Radio Physics Department, Tomsk State University, Tomsk, Russia
e-mail: yvlp@mail.tsu.ru



S. M. Goodnick et al. (eds.), *Semiconductor Nanotechnology*, Nanostructure Science and Technology, https://doi.org/10.1007/978-3-319-91896-9_8

inhomogeneities, so that the diffraction effects and effects of multiple interactions are of great importance. For that reason, this form of tomography is sometimes called diffraction tomography. Without dwelling on all the different methods and approaches of diffraction tomography [1–33] that are currently available, let us focus instead on active location (detection) wave tomography, which is of vital importance, e.g., for security systems.

Radio wave systems are preferable in the development of contactless detection devices for a variety of reasons. In the first place, radio waves are practically harmless to human health. This is their crucial difference from ionizing X-rays. Second, the potential range of application of these systems is quite wide: in crowded public places, in special-forces raids for detection and tracking of people hiding behind walls, detection of injured persons after emergency events, etc. There is also a great demand for contactless and computer-aided systems for quality control in building construction, timber processing, and other industries.

The variety of physical processes involved in radio wave detection, taking place in natural and simulated complex environments and involving complex objects, underlies the complexity of the mathematical descriptions of such processes and the urgency of solutions of the tomography problem as well.

Figures 8.1 and 8.2 show the known microwave passenger inspection systems. The screening of one person in this way takes 10–30 s, and the resolution obtained is

**Fig. 8.1** The model of SafeScout 100 Scanner

Fig. 8.2 The model of Smiths Heimann

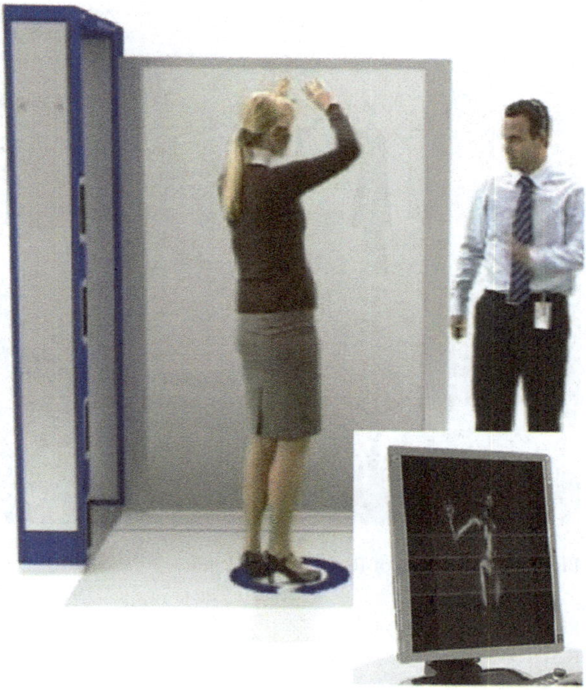

no better than 1 cm. The complexity of this system, substantial costs, and low resolution are its main disadvantages.

The use of ultra-wideband (UWB) systems and signals is considered to be the most promising approach from the standpoint of applications. The development of a radio-detection UWB tomograph requires the solution of a number of problems. Available commercial radio-detection include the Israeli scanner Raptor-1600 (Fig. 8.3), which is able to monitor the movement of large objects, e.g., a person moving behind a wall, and the Russian radio wave UWB scanner "RADIOVISION" (Fig. 8.4) which operates in real time at a speed of 10 fps.

As we see the radio wave systems already exist. However, this does not exhaust all of the currently available possibilities. The development of an optimal antenna array configuration is, of necessity, the first step in this process. This development would include the relative placement of the receiving and transmitting antennas and a determination of their minimum number required for tomographic imaging. The solution of this problem is one of the objectives of ongoing research. It includes determination of the optimal sequence of radar measurements to provide the required data digitization in a short period of time. Moreover, considerable attention must be paid to the development of fast algorithms for real-time 3D image restoration of scanned objects based on radio sounding data.

The antenna array optimization problem alluded to above also arises from the need to minimize the number of antennas in order to reduce antenna array costs

Fig. 8.3 The model of Raptor-1600 Scanner

Fig. 8.4 The model of RADIOVISION

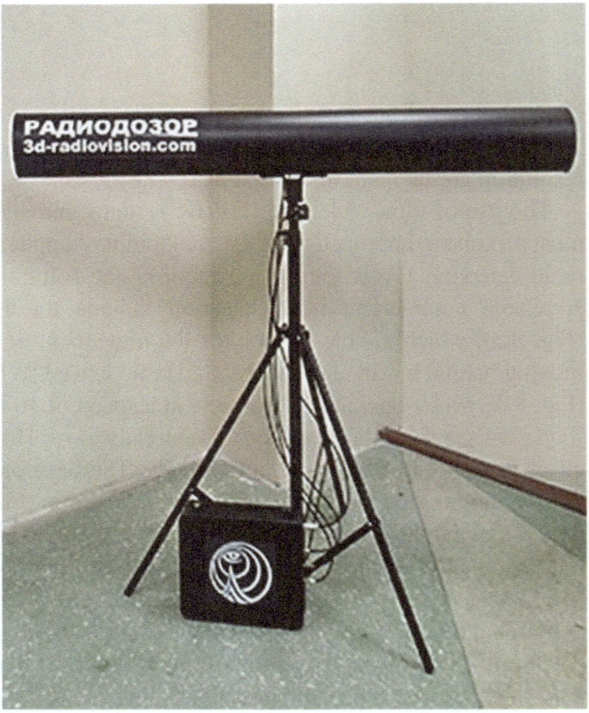

while holding the number of artifacts to a minimum. From the latter half of the twentieth century onward, several methods of antenna array optimization for radio astronomy and aircraft detection have been developed. The antenna array optimization problem is a natural outgrowth of the problem referred to as antenna array design with minimum redundancy. Accordingly, the antennas should be positioned in such a way as to effectively solve the problems that arise.

Over the past few decades, a great deal of research has been conducted, and field-proven results have been obtained in the far-field zone of narrowband antenna arrays. However, a subject of particular interest to the authors is antenna arrays for ultra-wideband radiation, which are designed to characterize objects located in the array focusing area. So far, this issue has not been investigated as there has not been any urgency to research the near-field zone of antenna arrays. In this context, the near-field zone is considered as the region of space at a distance on the order of the dimensions of the array. In other words, this refers to the Fresnel diffraction zone.

The next development of tomogram methods is the construction of images of hidden objects by using wave projections. In this context, the two following requirements come to the forefront: adequate restoration accuracy and real-time performance of the system. There is some contradiction here, which is formulated as "Fast is not always best." Reconstruction should be performed in real-time mode, for instance, in a minute. The task is complicated when it is impossible to obtain complete measurement data. This situation arises when the scanning is performed non-equidistantly, in motion, or when the object to be detected is hidden inside a building or under clothing near the body.

A phase information record is provided by radio-frequency holography, in which the result of interference of the background wave and the object wave is recorded. The wave projection record is in fact a radio-hologram record. From general considerations, it is clear that large aperture synthesis is the preferable method for post-processing these projections since it provides the highest spatial resolution. Currently there are a large number of variations of this method under continuous development [33].

Electromagnetic radiation interacts with electrophysical inhomogeneities in the propagation medium. In contrast to electromagnetic radiation, acoustic radiation interacts mainly with density inhomogeneities. In this context, using ultrasound for tomography of an inhomogeneous medium provides a wide range of additional possibilities, particularly for determining the type of material that the hidden items are made of.

The current state of research on remote ultrasonic sounding has revealed that practically all of these methods are based on contact measurements. The reason is that ultrasound is strongly attenuated by air. The only ultrasonic sounding system (sonar vision) which is actually efficient in air is bat echolocation. Some insects also generate ultrasound, but rather for scaring and masking from bats. The practical use of ultrasound in a liquid medium or through immersion liquids (e.g., in metal working, medicine, etc.) is well known. Preliminary studies hold out the hope for the effective use of ultrasound for purposes of near-field tomography in air, and our main focus areas include:

- Physical and mathematical models of an image reconstruction system for an inhomogeneous medium based on tomographic processing of multi-angle projection records of scattered radio wave and acoustic radiation
- Key elements of the modeling system
- Subsystems and elements of a multi-angle measuring tool
- Experimental measurement techniques
- Evaluation of potential and actual performance-based specifications for a detecting tomograph with different measurement and sounding schemes

The wide practical application of UWB tomography is still limited, on the one hand, by considerable engineering problems in generating and receiving UWB radiation and by complications, on the other hand, that arise when describing and interpreting simultaneously manifested physical phenomena of the interaction of the radiation with matter. These phenomena include multiple scattering, diffraction, wave interference, and absorption of ultra-wide-bandwidth radiation from arbitrarily placed and randomly oriented inhomogeneities of different size. A multitude of combinations of these effects hinders solution of the direct problem – a description of the integral effects of the wave disturbance. Solution of the inverse problem and reconstruction of the distribution of inhomogeneities in the tested volume are problematic under such conditions in any event. Researchers have no choice but to solve such inverse problems. These tasks are generally referred as ill-posed problems, which require the use of regularizing algorithms. The most stable inverse problems are the simplest ones, which account for the dominant mechanisms of the interaction of waves with the propagation media and make it possible to single out (identify) such mechanisms.

Inverse problems are of crucial importance for applications such as sounding the optically opaque media. When radio-frequency radiation penetrates into such a medium, analysis of the transmitted and scattered fields makes it possible to reconstruct its internal structure. This internal structure consists in the spatial distribution of the permittivity. Steep permittivity gradients are typical for interfaces and for immersed objects. A typical example is searching for hidden archaeological graves, underground cables, or antipersonnel landmines and ground covers as well as detection and identification of prohibited items in stowed luggage and in hand-carried items. Problems of this sort are not simple, but a great number of efficient solutions have been developed for them based on radiation focusing. With regard to a radiopaque object, radiation hardly penetrates it, so the solution of the inverse problem in this case reduces to reconstructing the shape of the object on the basis of an analysis of scattered (reflected) radiation.

Interesting directions for the future development of radiotomography include techniques for implementing incoherent radiation and low-frequency magnetic fields. The spectrum of radio tomography applications is extremely wide.

## 8.2 Focusing Wave

Focusing is the main tool that allows you to visualize probed heterogeneities using both pulsed and monochromatic radiation. Let us elucidate the essence of the proposed solution in the particular case of a homogeneous background medium in the single scattering approximation

$$E(\boldsymbol{\rho}_0, f) = k_1^2 \iiint\limits_{V_1} \Delta\varepsilon(\boldsymbol{\rho}_1, z_1) G_0{}^2(\boldsymbol{\rho}_1 - \boldsymbol{\rho}_0, z_1)\, d^2\boldsymbol{\rho}_1 dz_1. \tag{8.1}$$

Here $k_1 = k = 2\pi f/c$ is the wave number corresponding to the background medium. Green's function is written in this case as a spherical wave field:

$$G_0(\boldsymbol{\rho}_1 - \boldsymbol{\rho}_0, z_1) = \exp\left\{ ik\sqrt{(\boldsymbol{\rho}_1 - \boldsymbol{\rho}_0)^2 + z_1^2} \right\} / 4\pi\sqrt{(\boldsymbol{\rho}_1 - \boldsymbol{\rho}_0)^2 + z_1^2}.$$

If we differentiate Eq. (8.1), we can write

$$\frac{d}{dk}\left\{ \frac{E(\boldsymbol{\rho}_0, f)}{k^2} \right\} = \int\int\limits_{V_1}\int \Delta\varepsilon(\boldsymbol{\rho}_1, z_1) G_2(\boldsymbol{\rho}_1 - \boldsymbol{\rho}_0, z_1) d^2\boldsymbol{\rho}_1 dz_1,$$

where

$$G_2(\boldsymbol{\rho}_1 - \boldsymbol{\rho}_0, z_1) \equiv \exp\left\{ i2k\sqrt{(\boldsymbol{\rho}_1 - \boldsymbol{\rho}_0)^2 + z_1^2} \right\} / 4\pi\sqrt{(\boldsymbol{\rho}_1 - \boldsymbol{\rho}_0)^2 + z_1^2}.$$

This function admits a plane-wave decomposition of the spectrum (Weyl's formula):

$$G_2(\boldsymbol{\rho}_1 - \boldsymbol{\rho}_0, z_1) = \frac{i}{(2\pi)^2} \int\int \frac{\exp\{i(\boldsymbol{\kappa}_\perp(\boldsymbol{\rho}_1 - \boldsymbol{\rho}_0) + \boldsymbol{\kappa}_z z_1)\}}{2\boldsymbol{\kappa}_z} (d^2\boldsymbol{\kappa}_\perp),$$

where $\boldsymbol{\kappa}_z = \sqrt{(2k)^2 - \boldsymbol{\kappa}_\perp^2}$. Taking this representation into account, we can write

$$E_1(\boldsymbol{\rho}_\perp, f) \equiv \int\int \exp\{i\mathbf{u}_{\perp 0}\boldsymbol{\rho}\} \frac{d}{dk}\left\{ \frac{E(\boldsymbol{\rho}_0, f)}{k^2} \right\} d^2{}_0 =$$
$$\iint\limits_{V_1}\int \Delta\varepsilon(\boldsymbol{\rho}_1, z_1) \frac{\exp\{i(\boldsymbol{\rho}_{\perp 1} + u_z z_1)\}}{2i\mathbf{u}_z} d^2\boldsymbol{\rho}_1 dz_1, \tag{8.2}$$

where $\mathbf{u}_z = \sqrt{(2k)^2 - \mathbf{u}_\perp^2}$. This expression implies that the spatial frequency spectrum of the inhomogeneities coincides with the obtained expression for $E_1(\mathbf{u}_\perp, f)$ to within a factor:

$$\Delta\varepsilon(\mathbf{u}_\perp, \mathbf{u}_z) \equiv \iiint\limits_{V_1} \Delta\varepsilon(\boldsymbol{\rho}_1, z_1)\exp\{i(\mathbf{u}_\perp \boldsymbol{\rho}_1 + \mathbf{u}_z z_1)\} d^2\boldsymbol{\rho}_1 dz_1$$
$$= 2i\mathbf{u}_z E_1(\mathbf{u}_\perp, f). \tag{8.3}$$

In the final count, the three-dimensional inverse Fourier transform remains to be performed to recover the spatial distribution of the inhomogeneities. There is only one calculational peculiarity here if expression (8.2) is used, and that is the need to transform from time frequencies $f$ to the corresponding spatial frequencies

$$u_z = \sqrt{(2k)^2 - \mathbf{u}_\perp^2} = \sqrt{(4\pi f/c)^2 - \mathbf{u}_\perp^2},$$

which is realized by interpolation. Expression (8.3) realizes the idea of the focusing method at all distances.

Here and below, the problem of large aperture synthesis with radiation focusing is called radio wave tomosynthesis. Equation (8.3) in this case is a generalization of the Stolt method [28, 33], and it takes into account the difference between the functions $G_0(\boldsymbol{\rho}_1 - \boldsymbol{\rho}_0, z_1)$ and $G_2(\boldsymbol{\rho}_1 - \boldsymbol{\rho}_0, z_1)$. It should be noted that this substitution acts as a selection of the window function and does not affect the focusing procedure.

The proposed method is applicable in the case of a multilayer medium. It suffices, for this purpose, to replace the exponential factor $\exp(iu_z z_1)$ by the combined factor

$$\exp\left\{ i\sum_j u_{zj} z_j + iu_z z_1 \right\},$$

which accounts for the phase difference for all previous layers with depths $z_j$ and refractive indices $n_j$. The normal projection of the wave number in each of the layers is calculated as

$$u_{zj} \equiv \sqrt{(2kn_j)^2 - \mathbf{u}_\perp^2}.$$

As an experimental test of this approach, a model medium was constructed from three foam-concrete blocks with equal depth of 10 cm (Fig. 8.5a). A thin test layer made from five aluminum foil strips with equal width of 2 cm (Fig. 8.5b) was placed between the second and third blocks. The aluminum test layer is indicated by an arrow. The measured refractive index of the foam-concrete was $n = 1.44$. A radio image of the test layer obtained by the proposed method is shown in Fig. 8.5b as a gray-scale plot. Sounding was performed using UWB pulses with a pulse duration of 200 ps.

The radio image reproduces the actual size and positions of the aluminum strips with a resolution of at least 2 cm. If the background medium is inhomogeneous, the resolution will be a bit worse.

The fundamental idea of radio wave tomosynthesis is to develop a focusing effect as a result of partial wave interference. The connection between radio wave

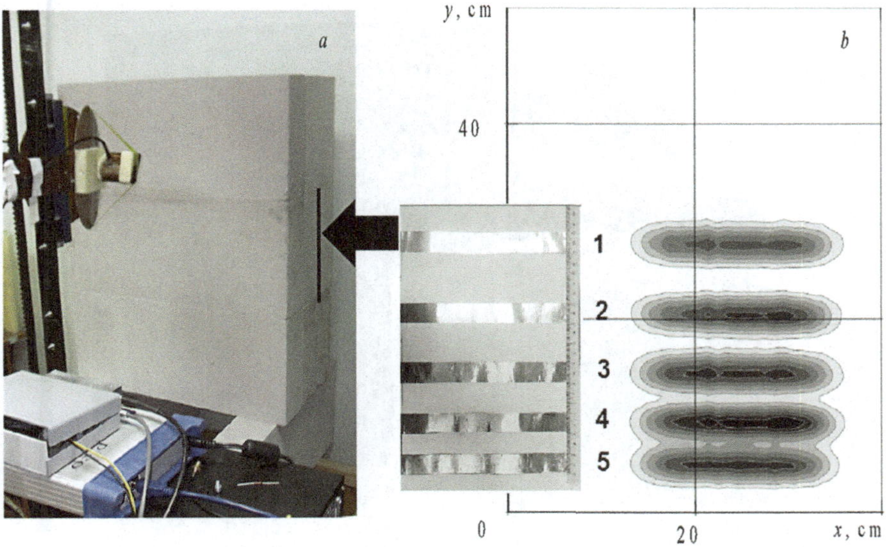


**Fig. 8.5** Experiment to validate the group focusing approach: (**a**) a wall made of foam, concrete blocks; (**b**) test object inside the wall and its radio image

tomosynthesis and multidimensional matched filtering and other methods has been pointed out.

It has been demonstrated that the focusing effect, which is the basis of wave tomography, significantly reduces the effect of multiple interactions and amplifies the role of the dominant mechanisms of the radiation-matter interaction.

The emphasis has been upon location sounding, where the radiation and receiving points are located in the same half-space. Many of the algorithms proposed for the solution of the inverse problem reduce to fast algorithms and admit of real-time operation.

Most of the solutions considered have been borne out by numerical simulation and real experiments.

## 8.3 Experimental Examples

### *8.3.1 Antennas and Signals*

The basic setup for UWB radio wave sounding includes a scanner, an XY positioner (Fig. 8.6), a scanner control unit, a TMR8140 stroboscopic oscilloscope (Fig. 8.7), a personal computer, and a transceiver module (Fig. 8.8).

The transceiver module is worthy of note. The original design of the UWB antenna is implemented in this device (Fig. 8.9). The antenna is constructed two electric antennas (a TEM horn and an asymmetrical dipole) and a magnetic antenna

**Fig. 8.6** Scanner for UWB measurements

**Fig. 8.7** TMR8140 stroboscopic oscilloscope and bipolar pulse generator with a duration of 0.1 ns

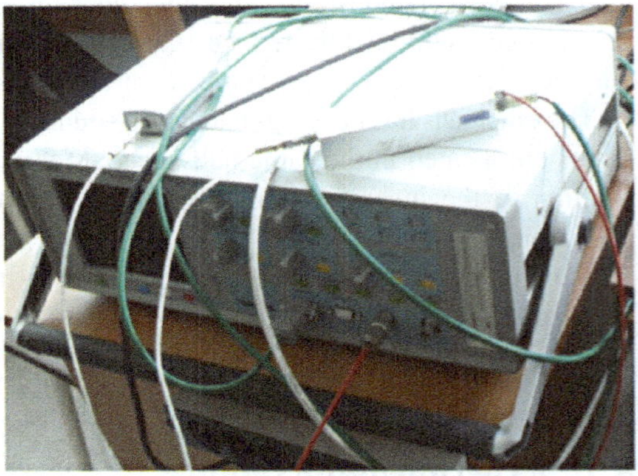

of spiral shape. The overlap of the near-field zones of these unlike (electric and magnetic) antennas makes it possible to considerably reduce the presence of reactive fields and extend the bandwidth as a result.

The displacement of the antenna phase center for the frequency spectrum does not exceed its diameter. The design of the antenna was developed by Associate Professor J. I. Buyanov.

Fig. 8.8 Transceiver module

**Fig. 8.9** UWB antennas of snail type

The frequency characteristic of the smallest of the antennas is shown in Fig. 8.10. The bandwidth of this antenna is 1–18 GHz. The voltage standing wave ratio (VSWR), which is a specification used to rate antennas, does not exceed 2.

The directivity diagram of the antenna, by virtue of its small size, is quite wide in both the azimuthal and the elevation plane (30–70°). Directivity diagrams (DDs) of the antenna, measured at two frequencies (3 GHz and 9 GHz), are shown in Fig. 8.11.

Signals with durations of 100, 200, and 1500 ps were analyzed for radio sounding. Equipment from TRIM, Ltd. – Research and Production Enterprise (St. Petersburg, Russia) (TRIM Ultrawideband Measurement Systems) – was used to generate and receive the UWB signals. The pulse repetition rate of the strobe was 100 kHz. Analysis of signal forms and spectra (Fig. 8.12) revealed that pulses with durations of 100 and 200 ps were most relevant for radio tomography. These signals in particular should provide proper spatial resolution, and their spectra are well-matched to the chosen design of the transceiver antenna.

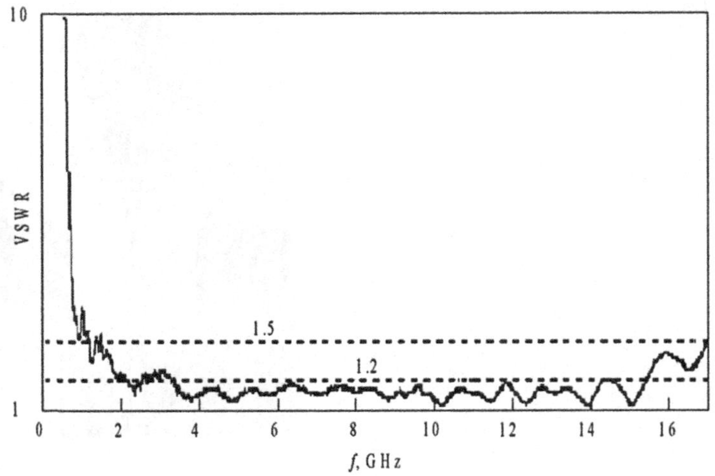


**Fig. 8.10** VSWR frequency dependence of snail-type antennas

A typical signal with a duration of 100 ps, reflected from the test object, is shown in Fig. 8.13. The signal is of indented form, and the central peak is delayed by the travel time from the object to the antenna module.

Figure 8.14 displays a sample raster image of recorded signals from scanning the test object. The line-by-line rasters presented here are set one above another. In this case, horizontal flyback of the scanner starts immediately after direct scanning of the previous line.

This considerably shortens the scanning time. It is important that the characteristic diffraction hyperbola, which indicates the localization of the object, is distinctly visible at every step.

The scanner control unit contains a USB-to-COM converter based on the FTDI 245BM chip. This makes it possible to transform low-current USB command signals into eight binary output signals, which after being routed through matching amplifiers are directed to the stepper motor control of a two-coordinate scanner. There are four windings in each stepper motor, where the current in each of them assigns the motor state. The control unit makes it possible to position the scanner antenna module along two axes independently. In general, the experimental setup allows the transceiver antenna module to be repositioned within an 84 $\times$ 84 cm square region in the plane with an accuracy of 2 mm and a speed of up to 3 cm/s.

Employing monostatic radiometry methods to solve problems of tomography in most cases supposes the use of a single antenna for both reception and transmission in a way that facilitates more accurate reconstruction of the radio wave image of the tested object, all other things being equal. In UWB tomography, the use of a single antenna is complicated by technical difficulties associated with isolation of the transmitting and receiving antenna paths. It proved to be impossible to find suitable directional couplers providing isolation of not less than 30 dB in the UW frequency band. Development of a fast UWB switch is another independent task.

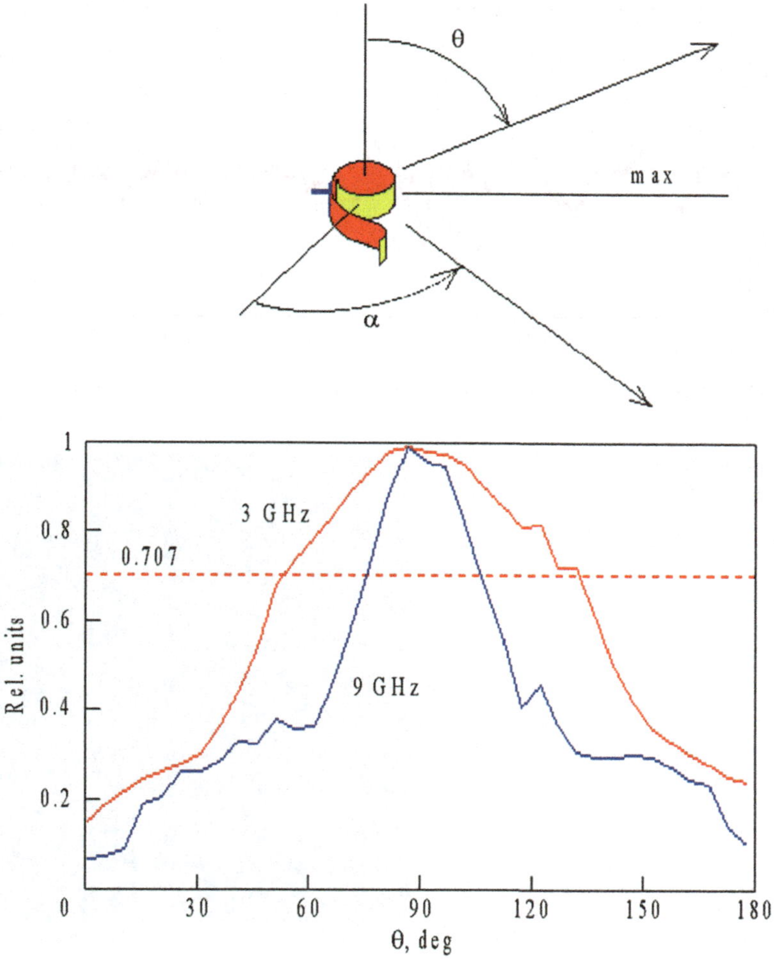


**Fig. 8.11** Azimuthal DD of a snail-type antenna

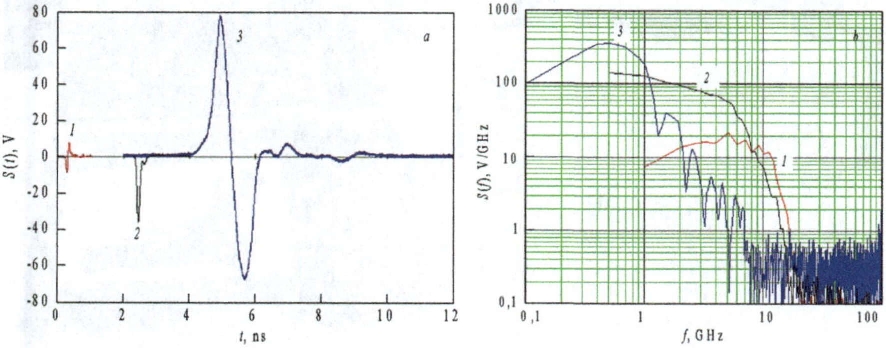


**Fig. 8.12** UWB pulses of different duration (**a**) used for sounding and their spectra (**b**): curve *1*–100 ps; curve *2*–200 ps; curve *3*–1500 ps

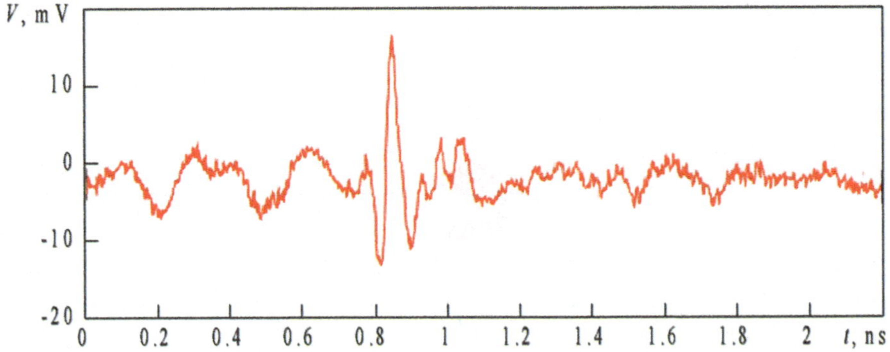


**Fig. 8.13** A UWB signal reflected from the test object

**Fig. 8.14** Radio wave scanning image of tested section

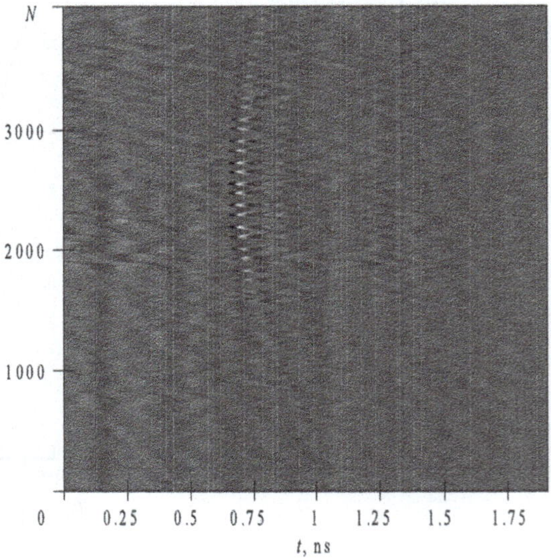


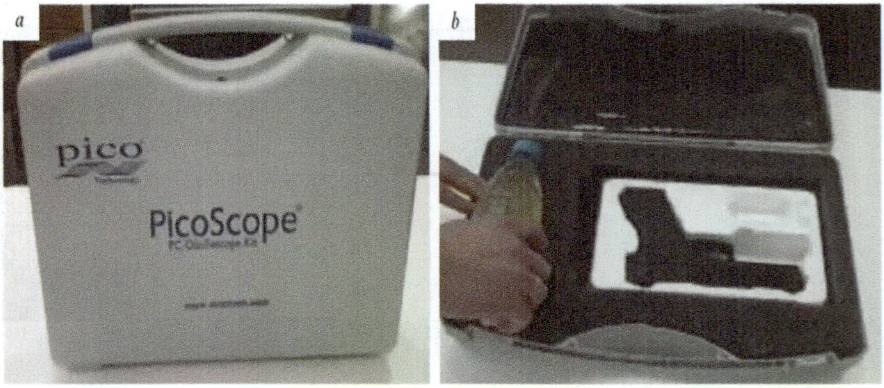


**Fig. 8.15** Plastic suitcase (**a**) and case contents (**b**)

It was experimentally determined that the antennas must be separated by a distance of not less than 20 cm to get isolation of 25–30 dB between the receiving and transmitting antennas; however, this is unacceptable in UWB tomography.

To increase the decoupling from the transmitting antenna to the receiving antenna, a metallic screen was used, whose shape and geometric dimensions were determined experimentally. It turned out that an acceptable result was achieved when the screen had a shape close to that of an ellipse with major and minor semiaxes of 9 and 5 cm, respectively. A photograph of the assembly of antennas with screen is shown in Fig. 8.9.

### *8.3.2 Different Various Objects and Their Radio Image*

UWB tomography is especially effective for luggage inspection. A plastic suitcase with a plastic gun and a bottle of water inside it is shown in Fig. 8.15. Figure 8.16 presents the tomography of this scene.

The experimental setup described above (see Fig. 8.7) was used for tomosynthesis of images of hidden objects in media with metallic inclusions. First off, a wooden carrying box reinforced with metal bands was sounded (Fig. 8.17). The dimensions of the box were 47 × 41 × 19 cm.

A metal-coated stepped triangle was placed in the box. Each step of the triangle was 5 cm in length. There was a 2 × 2 cm square hole in the center of the triangle (Fig. 8.18).

Scanning was performed by moving the transceiver antenna module in the *OXZ* plane with a step of 1 cm. The scanning area had dimensions 60 × 50 cm. The test

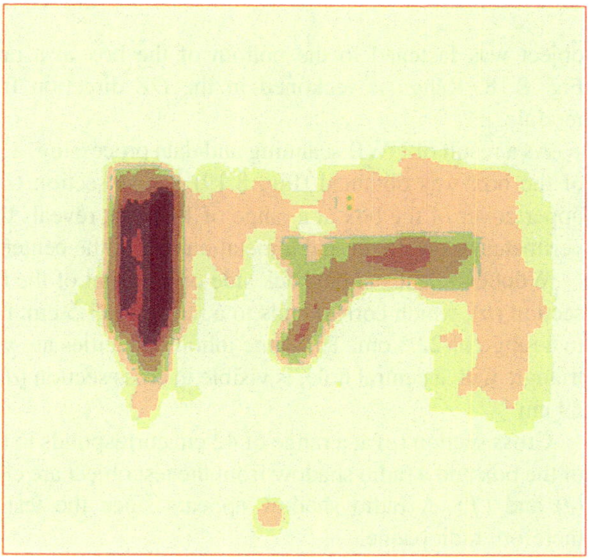

**Fig. 8.16** UWB tomogram of case contents

**Fig. 8.17** UWB scanner and carrying box

object was fastened to the bottom of the box at a range of 39 cm, as shown in Fig. 8.18. Range is reckoned in the *OZ* direction from the transceiver antenna module.

As a result of UWB scanning and data processing, a 3D tomogram of the contents of the box was obtained (Fig. 8.19). Cross section (*a*), which corresponds to the upper cover of the box at a range of 18.5 cm, reveals the reinforcing tape along the perimeter of the cover and a metal handle in the center of the cover.

Wooden latches at the back side of the front of the box can be observed in cross section (*b*), which corresponds to a range of 21.5 cm. Cross section (*c*) corresponds to a range of 22.5 cm. The same inhomogeneities are visible in it. The test object, a triangle with a central hole, is visible in cross section (*d*) corresponding to a range of 39 cm.

Cross section (*e*) at a range of 42 cm corresponds to the back of the box. The back of the box and a radio shadow from the test object are clearly visible in cross sections (*e*) and (*f*). A radio shadow appears since the test object is metal-coated and therefore radiopaque.

**Fig. 8.18** Test object inside the box

According to the results of the experiment, it may be concluded that implementation of tomosynthesis algorithms makes it possible to tune out interference caused by echo signals reflected from the reinforcing elements. The obtained tomogram allowed us to determine the depth of the test object with an accuracy better than 0.5 cm and visualize its shape as well.

Next, let us consider experimental results of visualization of an object placed in an open metal container. A metal safe with dimensions 64 × 48 × 41 cm was used for this purpose (see Fig. 8.20). The door of the safe was taken off its hinges. Figure 8.21 shows the safe from behind.

On completion of scanning, a tomogram was obtained which showed the distribution of inhomogeneities in the tested space. Radio images of cross sections at different ranges are presented in Fig. 8.22. Figure 8.22a presents a cross section corresponding to the upper cover of the safe at a range of 18.5 cm. The front frame of the safe stands out distinctly in the tomogram. The cross section corresponding to a range of 19 cm is shown in Fig. 8.22b. This tomogram demonstrates the sunken frame used to hang the door. The cross section corresponding to a range of 53 cm is shown in Fig. 8.22c. In this tomogram, the test object with central hole is distinctly visible. The cross section at a range of 62 cm is shown in Fig. 8.22d. This cross section images the back wall of the safe.

Based on the experimental results, it may be concluded that implementing tomosynthesis algorithms makes it possible to tune out interference caused by echo signals reflected from the walls and bottom of the metal container. The obtained

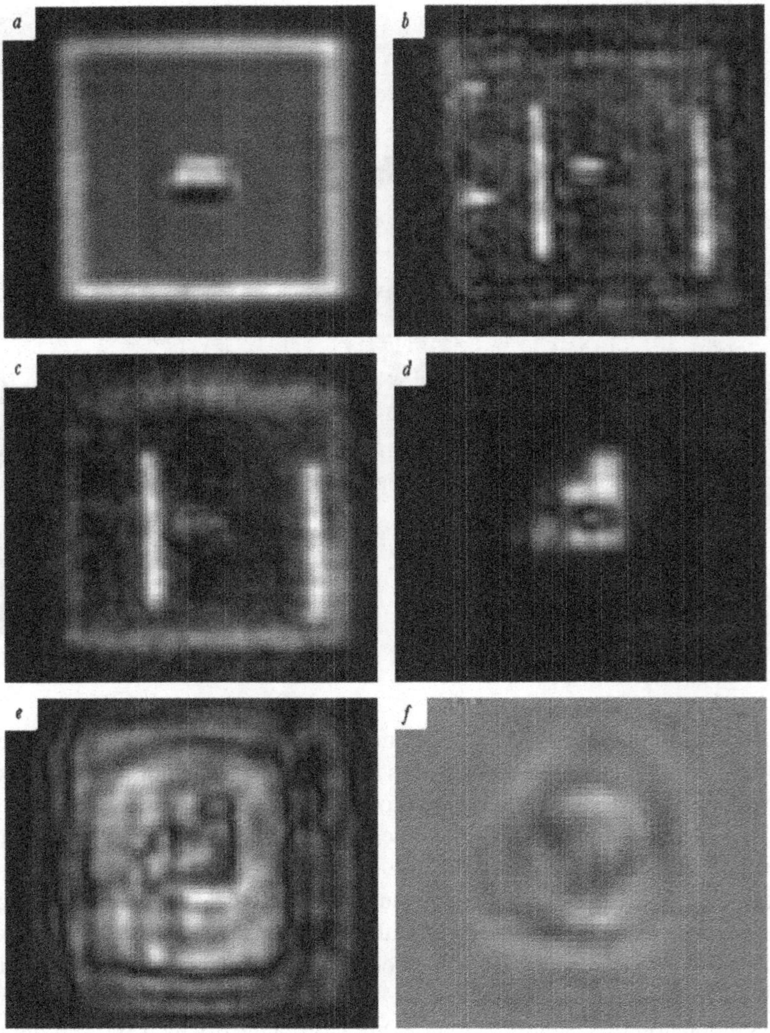


**Fig. 8.19** Radio wave tomogram of a wooden box (with a metal band along its edge) and its contents

tomogram allows the depth of test object to be determined and the shape of the object to be visualized as well.

Searching big trucks in order to prevent the transportation of prohibited articles such as weapons, explosives, etc. is a major problem at access control points (ACPs). There are special-purpose X-ray detectors for that purpose, but it is not possible to equip all of the critical ACPs owing to the high cost of the equipment. A relatively cheaper radio wave system for detecting and visualizing hidden objects can be implemented to search dump trucks and similar vehicles. It is depicted in Fig. 8.23.

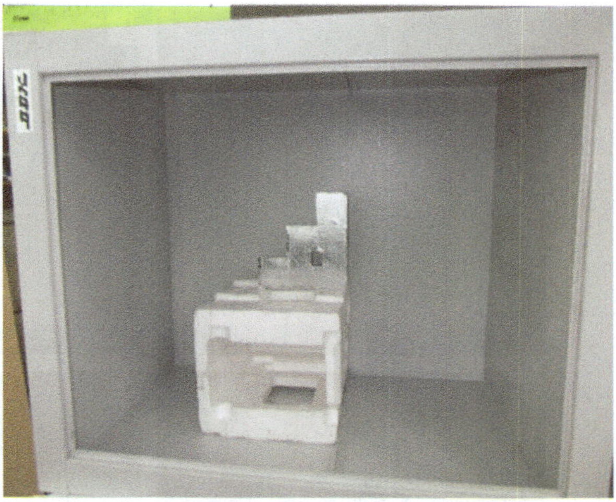

**Fig. 8.20** Test object inside the metal safe

**Fig. 8.21** Back view of the safe

The system is basically a clocked array of UWB antennas, which are switched by electromechanical switching devices. When a truck moves through this system at low speed, it will be possible to sound the truck body in order to detect hidden objects under the load (soil, sand, etc.). A prototype of this system is shown in Fig. 8.4.

The best possible resolution in the radio range can be obtained in the millimeter frequency range. This is the range where practically all American radio scanner prototypes work. It is an important fact that the resolution in this band is usually sufficient to recognize fine details of an object, while the penetration remains on a reasonable level.

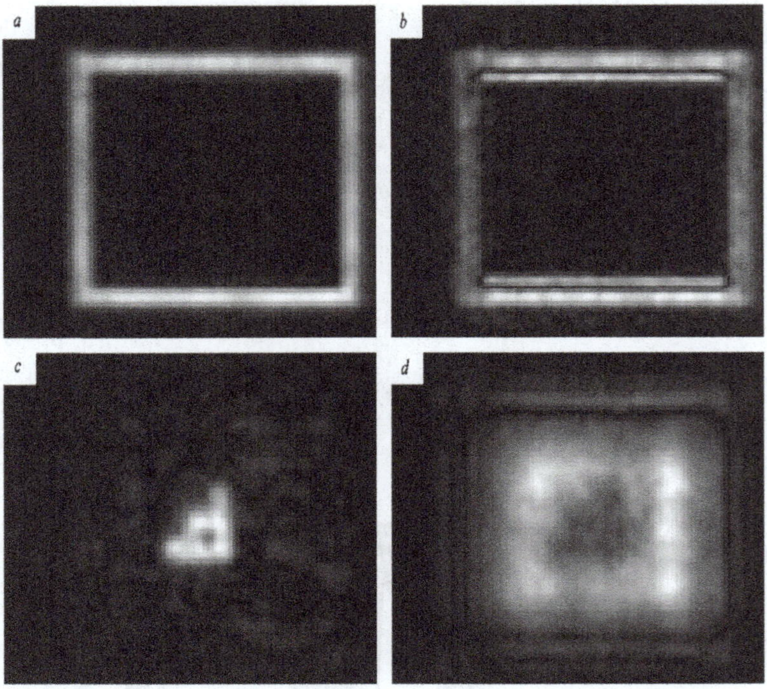


**Fig. 8.22** Ultrasonic tomogram of a metallic box and its contents

**Fig. 8.23** Advanced system for detecting and visualizing hidden objects at ACPs

### 8.3.3 Ultrasonic Vision

Electromagnetic radiation interacts with electrophysical inhomogeneities in the propagation medium. In contrast to electromagnetic radiation, acoustic radiation interacts mainly with density contrasts in the sounding medium. In this context, using ultrasound in the tomography of inhomogeneous media provides additional opportunities, in particular for detecting the type of material that the hidden objects are made of. Integration of radio and acoustic sounding provides opportunities, e.g., for the detection of explosives.

A search on remote ultrasonic sounding techniques reveals that most industrial methods are based on contact measurements. The principal reason for this is that ultrasound attenuation in air is quite high. Ultrasound is used in Parktronic systems for car parking and for control of industrial robots. The ultrasound system (echolocation) used by bats is perhaps the only ultrasonic ranging system actually effective in air (see Fig. 8.24).

Some insects also use ultrasound, but rather for active jamming of bats' sounding. There is no doubt in the efficiency of ultrasound for sounding in liquids and through immersion liquids. This applies in the fields of industrial metal working, medicine, and submarine echolocation. In nature this also pertains to dolphins and fish. However, ultrasound can be efficiently used for tomography in air in safety systems and in nondestructive testing. If the object is radiopaque, radiation can hardly penetrate it, and only acoustic radiation will allow the recovery of its internal structure.

From a mathematical perspective, acoustic wave propagation is similar to the propagation of electromagnetic waves. Both electromagnetic and acoustic waves are described by wave equations, and both give rise to reflection, scattering, and diffraction effects when the wave interacts with inhomogeneities in the medium. In this respect, all of the mathematical methods and results elaborated in previous chapters are applicable for ultrasound.

The simplest setup for our experiments in ultrasound tomography is shown in Fig. 8.25. Standard piezoceramic transmitters and ultrasound receivers with a

**Fig. 8.24** Bats are effective ultrasonic stations in air

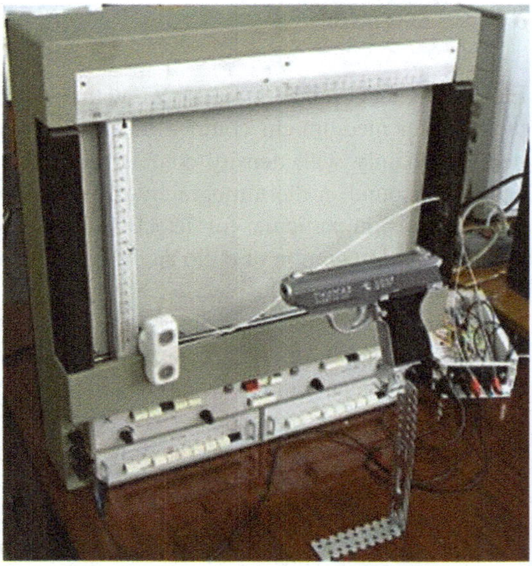

**Fig. 8.25** Experimental setup for ultrasonic tomography

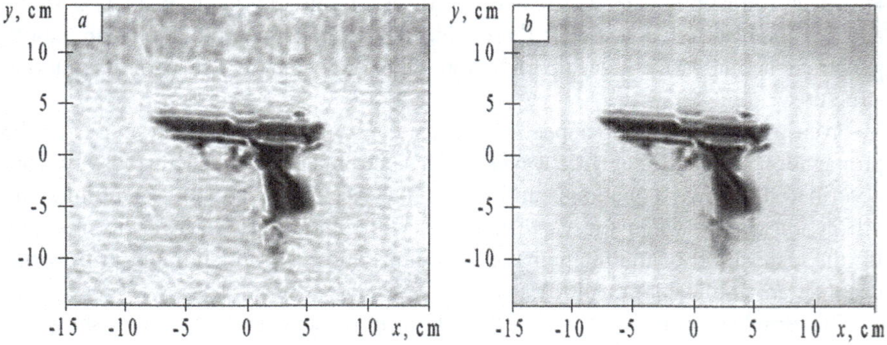


**Fig. 8.26** Recovered image of the object (tomogram): at a frequency of 40 kHz (**a**), within the frequency band 37–43 kHz (**b**)

resonant frequency of 40 kHz are set on an x-y recorder, and a computer sound card can be employed as a signal generator and receiver. A Creative Audigy SE PCI SB0570 sound card enables signal generation and recording with a sample rate up to 96 kHz. An MA40S4R ultrasonic transmitter was used as well as an EM9767 electret microphone. The object of sounding (a toy gun) was placed at a certain distance from the transmitter. Scanning was performed through a simple computer-aided controller.

The restored image of the test object is presented in Fig. 8.26. The image of a gun can be clearly distinguished in the result of single-frequency sounding; however, there are minor artifacts. The number of artifacts was markedly decreased when the

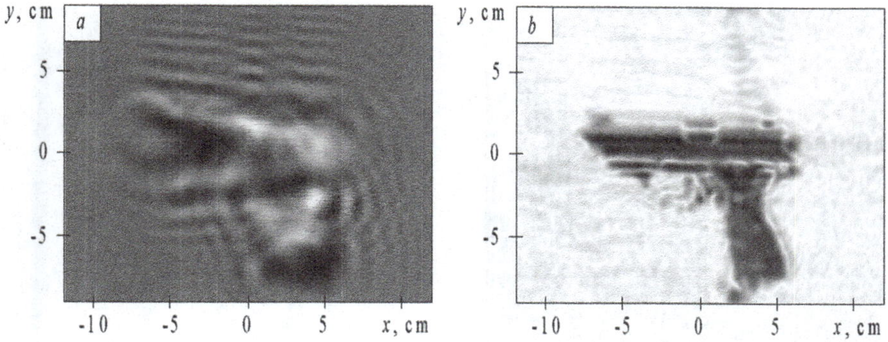


**Fig. 8.27** Plastic gun in air at a distance of 10 cm: measured data (**a**) and result of focusing (**b**)

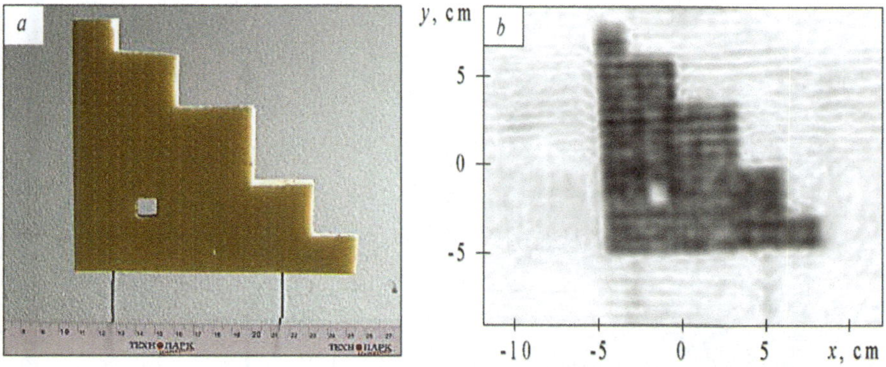


**Fig. 8.28** Stepped triangle in air at a distance of 10 cm: (**a**) photograph and (**b**) the result of ultrasonic imaging

sounding was performed at all frequencies within the frequency band 37–43 kHz. It should be noted that in air a wavelength of 8.3 mm corresponds to a frequency of 40 kHz, which makes it possible to distinguish fine details in the image.

The resolution of details improves as the distance to the object is decreased. This is because the condition of Fresnel diffraction is being met more precisely (Fig. 8.27).

The reported results confirm the efficacy of the use of ultrasound in location (detection) tomography of small-sized objects. Furthermore, no special permit is required to use ultrasound, so it can be used for covert surveillance.

The result of sounding of a test object in the form of a stepped triangle with a hole is displayed in Fig. 8.28. The resolution in this case is close to the radiating wavelength and even higher.

Objects can be visualized which are hidden behind opaque but acoustically transparent screens (Fig. 8.29). The image is blurred of course, but it remains recognizable. The resolution could be improved significantly with the use of multiple frequencies.

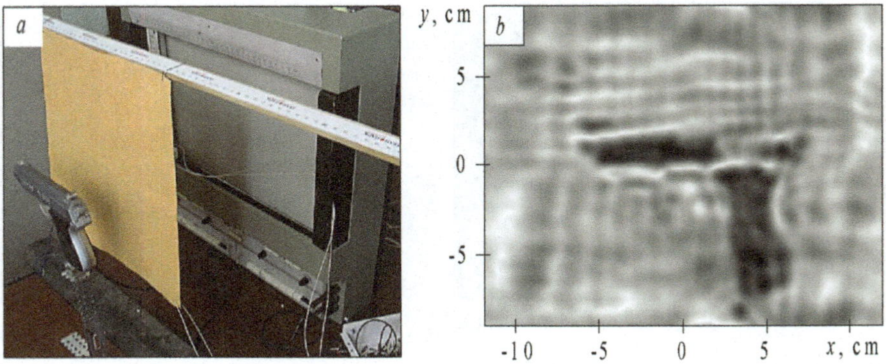


**Fig. 8.29** Plastic gun behind a curtain at a distance of 10 cm: (**a**) photograph of experimental setup and (**b**) the result of focusing

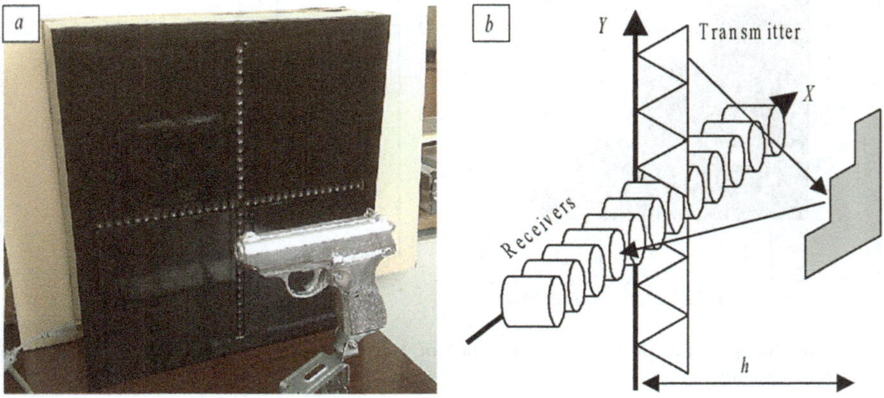


**Fig. 8.30** Sonar for contactless ultrasonic imaging: external view (**a**), configuration scheme (**b**)

Mechanical scanning of an acoustic field requires up to several tens of minutes, and the object must remain motionless, which is unattainable, e.g., when screening people. Thus, it would be interesting to consider how sounding could be speeded up by electronic switching of the transmit/receive arrays. In the view of the authors, cross-shaped sounding systems are of interest. A demo setup of a sonar was therefore developed that included 32 ultrasonic transmitters and 32 receivers arranged in the form of a cross, situated 1 cm apart from each other (Fig. 8.30).

Measurements were carried out in the clocked mode. The time for a complete measurement, data processing, and visualization of the results on a standard computer does not exceed 2 s, which is acceptable for many applications. An image obtained using the sonar demo setup is presented in Fig. 8.31.

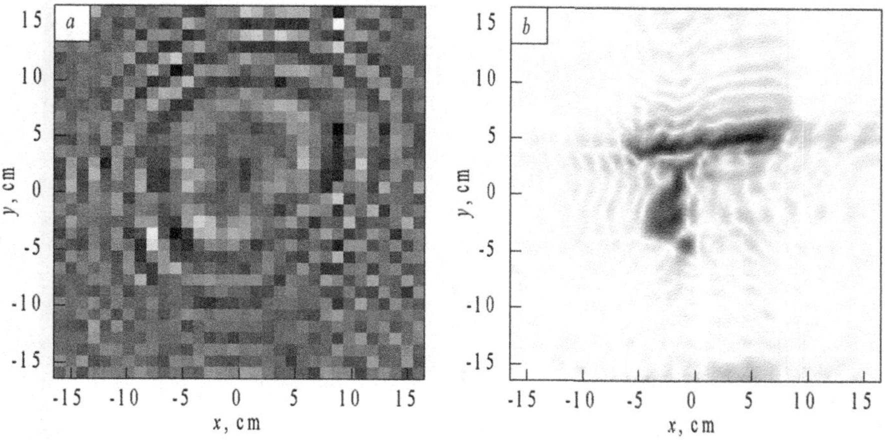


**Fig. 8.31** Testing results of sonar demo setup: measured field (**a**), recovered image (**b**)

## 8.3.4 *Integration of Radio and Ultrasonic Vision*

Ultrasonic radiation mostly is scattered at the inhomogeneities of the density ρ. In remote sounding, it provides the information about the material an inhomogeneity is made of. Using electromagnetic waves in sounding makes it possible to monitor the electrophysical properties of matter. The integration of electromagnetic and acoustic waves for sounding enables us to distinguish solid and liquid media. The following empirical relation is true for solids:

$$\rho = 2(n - 1), \text{or} \quad n = [\rho/2 + 1],$$

where ρ is the density and $n = \sqrt{\varepsilon}$ is the refractive index of the material. A plot of the refractive index versus the density is shown in Fig. 8.32, where the sloped line plots the above dependence.

Note that that TNT, which is distinctly different from other materials, takes a very low position in the diagram. The position of fluoroplastic (Teflon) is close to that of TNT. This means that Teflon can be used to simulate TNT in the laboratory; in real practice, it is necessary to use reconstruction of the shape of such objects to distinguish these materials.

A series of experiments with different objects was carried out (Fig. 8.32). The ultrasound frequency was tuned within the band of 35–45 kHz, and the duration of the UWB pulses was 200 ps. It was established that the images of metallic objects in radio wave tomography have more contrast than in ultrasonic tomography. Apparently, the reason for this is that a part of the energy is spent to excite secondary waves in the metal and behind it.

On the contrary, radio waves have weaker reflection from dielectrics and penetrate into dielectric objects. This was clearly demonstrated in experiments with a metallized grid (the reflection coefficient of radio waves was 0.9, whereas for

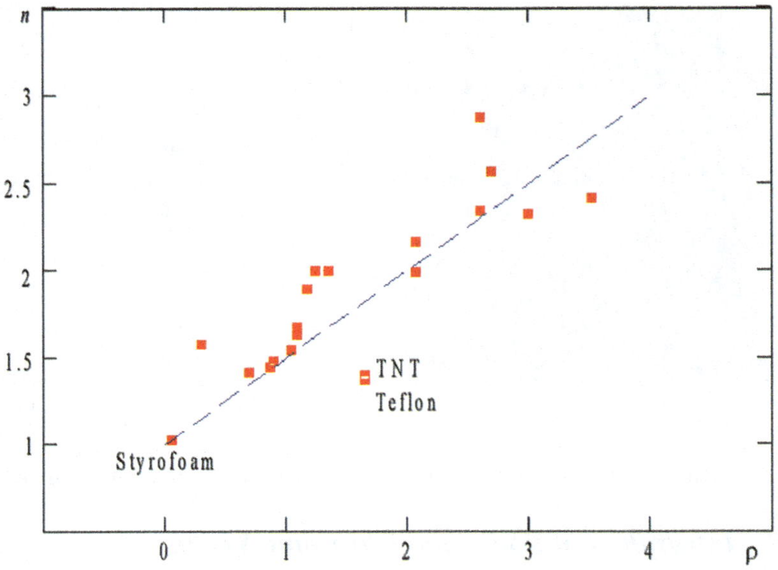


**Fig. 8.32** Refractive index – density diagram of several solid materials

ultrasonic waves it was 0.1) and with a foam-plastic object (where the radio waves did not reflect and the reflection coefficient of ultrasound was 0.7). Ultrasound provided higher resolution (of 2–3 mm) than the radio waves (1–2 cm) at identical distances (<1.2 m) to the test object.

The reason for this is the difference between the average operating wavelength for ultrasound (8 mm) and for a radio wave (2 cm). It is important here that the Fresnel zone (the focusing region) for ultrasound has about twice the diameter as for radio waves given the same size of synthetic aperture.

Overlaying the radio wave image and the ultrasonic image of the test object in a false-color combined image makes it possible to judge the material of the test object without additional processing (Fig. 8.33). A similar effect is observed in an element-by-element multiplication of images with a significant increase of the resolution as well.

The combined use of radio waves and ultrasound enables an increase in the information content of the obtained images both in resolution and in the possibility of identifying the material that the sounded object is made of. Radio waves provide penetrating power, for example, under clothing, and ultrasound increases the accuracy of identification of the material of the hidden object. A complete discovery of the possibilities of this approach would require individual study of a wide array of materials using the neural network method.

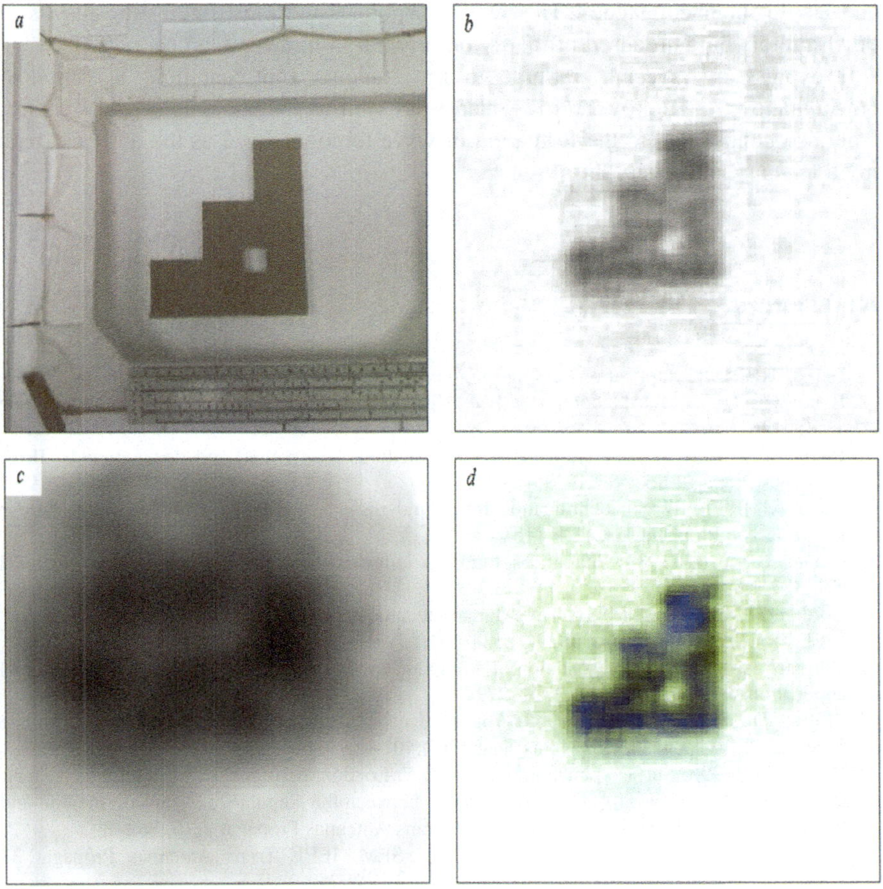

**Fig. 8.33** Image of a stepped triangle made of plasterboard behind a metallized grid: (**a**) photograph, (**b**) ultrasonic image, (**c**) UWB image, (**d**) image from integrated tomography

## 8.4 Conclusion

Here we have considered new methods of wave tomography. All methods are considered on the basis of a single approach, called tomosynthesis of radio waves. The basis is the idea of creating the effect of focusing radiation waves as a result of interference of a set of partial waves. The connection of the method of wave tomosynthesis with multidimensional sequential filtration and other methods is noted.

A large body of experimental data and their interpretation was obtained within the framework of international collaboration with leading scientists from Magdeburg University (Magdeburg, Germany), the Institute for Non-Destructive Testing (Saarbrücken, Germany), and Tohoku University (Sendai, Japan). I am deeply indebted to Professors A. Omar, M. Kroning, and M. Sato for their collaboration and also to their colleagues. This text incorporates the most interesting results of other authors but is focused primarily on original in-house research.

I express my special gratitude to my students and coauthors S. Shipilov, D. Sukhanov, A. Klokov, and R. Satarov, as well as the students of my students, who also contributed to the formation of wave tomography. I as the author extend my appreciation to everyone.

## References

1. Barrett HH (1982) Optical processing in radon space. Opt Lett (7, 6):248–250
2. Baum CE (1973) Introduction to SEM. IEEE Group Antennas Propagat Int Symp Dig 11:459–462
3. Baum CE (1986) The singularity expansion method: background and developments. IEEE Antennas Propagat Soc Newsl 28(4):14–23
4. Berni AJ (1975) Target identification by natural resonance estimation. IEEE Trans Aerosp Electron Syst Propagat 11(2):147–154
5. Chiang B-S et al (1983) Spatial resolution in industrial tomography. IEEE Trans Nucl Sci NS-30(2):1671–1676
6. Cordaro JT, Davis WA (1981) Time-domain techniques in the singularity expansion method. IEEE Trans Antennas Propagat 29(3):534–538
7. Cormack AM (1963) Representation of a function by its line integrals, with some radiological application. J Appl Phys 34(9):2722–2727
8. Moffatt DL, Kennaugh EM (1965) The axial echo area of a perfectly conducting prolate spheroid. IEEE Trans Antennas Propagat 13(3):401–409
9. Daniels DJ (1996) Surface-penetrating radar. IEE, London, p 300p
10. Dolph CL, Cho SK (1980) On the relationship between the singularity expansion method and the mathematical theory of scattering. IEEE Trans Antennas Propagat 28(6):888–897
11. Felsen LB (1985) Comments on early time SEM. IEEE Trans Antennas Propagat 33 (1):118–119
12. Gaunaurd GC, Uberall H, Nagl A, Subrahmanyam JV (1982) SEM poles and creeping-wave transients on conductors of simple shapes. IEEE Int Antennas Propagat Symp 20:16–17
13. Joaquim F-G (2002) A novel 3-D subsurface radar imaging technique. IEEE Trans Geosci Remote Sens 40(2):443–452
14. Kennaugh EM, Cosgriff RL (1958, March). The use of impulse response in electromagnetic scattering problems. IRE National convention record, pp. 72–77
15. Marin L (1974) Natural-mode representation of transient scattering from rotationally symmetric bodies. IEEE Trans Antennas Propagat 22(2):266–274
16. Mittra R, Pearson LW (1978) A variational method for efficient determination of SEM poles. IEEE Trans Antennas Propagat 26(2):354–358
17. Moffatt DL, Mains RK (1975) Detection and discrimination of radar targets. IEEE Trans Antennas Propagat 23(3):358–367
18. Mohan R, Rothenberg L, Reinstien L, Ling CC (1995) Imaging in three-dimensional conformal radiation therapy. Int J Imaging Syst Technol 6(1):14–32
19. Morgan MA (1984) Singularity expansion representations of fields and currents in transient scattering. IEEE Trans Antennas Propagat 32(5):466–473
20. Munk WH (1979) Oceanic acoustic tomography: a scheme for large monitoring. Deep-Sea Res 26(2A):123–161
21. Nagl A, Uberall H (1982) Transient radar scattering analyzed in terms of creeping waves. IEEE Int Antennas Propagat Symp Dig 20:18–20

22. Pearson LW (1984) A note on the representation of scattered fields as a singularity expansion. IEEE Trans Antennas Propagat 32(5):520–524
23. Pearson LW, Roberson DR (1980) The extraction of the singularity expansion description of a scatterer from sampled transient surface current response. IEEE Trans Antennas Propagat 28 (2):182–190
24. Ramm AG (1980) Theoretical and practical aspects of singularity and cexpansion methods. IEEE Trans Antennas Propagat 28(6):897–901
25. Richards MA (1994) SEM representation of the early and late time fields scattered from wire targets. IEEE Trans Antennas Propagat 42(4):564–566
26. Sahiner B, Yagle AE (1996) Iterative inversion of the radon transform using image-adaptive wavelet constraints to improve image reconstruction. IEEE Eng Med Biol 8(7):395–397
27. Sarkar TK, Weiner DD, Jain VK, Diant SA (1982) Impulse response determination in the time domain – theory. IEEE Trans Antennas Propagat 30(4):657–663
28. Stolt RH (1978) Migration by Fourier transform. Geophysics 43(1):23–48
29. Swith WE, Barrett HH (1983) Radon transform and band width compression. Opt Lett 8 (7):395–397
30. Tesche FM (1973) On the analysis of scattering and antenna problems using the singularity expansion technique. IEEE Trans Antennas Propagat 21(1):53–62
31. Tseng F, Sarkar TK (1984) Experimental determination of resonant frequencies by transient scattering from conducting spheres and cylinders. IEEE Trans Antennas Propagat 32 (9):914–918
32. VanBlaricum ML, Mittra R (1975) A technique for extracting the poles and residues of a system directly from its transient response. IEEE Trans Antennas Propagat 23(6):777–781
33. Yakubov VP, Shipilov SE, Sukhanov DYa, Klokov AV (2017) Wave tomography. In: Yakubov VP (ed). Scientific Technology Publishing House, Tomsk, 248p

# Index




S. M. Goodnick et al. (eds.), *Semiconductor Nanotechnology*, Nanostructure Science and Technology, https://doi.org/10.1007/978-3-319-91896-9

## D



## E



## F